COMPUTER MATHEMATICS

NANKAI SERIES IN PURE, APPLIED MATHEMATICS AND THEORETICAL PHYSICS

Published:

Vol. 1 Probability and Statistics
eds. Z. P. Jiang, S. J. Yan, P. Cheng and R. Wu

Vol. 2 Nonlinear Anaysis and Microlocal Analysis
eds. K. C. Chung, Y. M. Huang and T. T. Li

Vol. 3 Algebraic Geometry and Algebraic Number Theory
eds. K. Q. Feng and K. Z. Li

Vol. 4 Dynamical Systems
eds. S. T. Liao, Y. Q. Ye and T. R. Ding

Vol. 5 Computer Mathematics
eds. W.-T. Wu and G.-D. Hu

Vol. 6 Progress in Nonlinear Analysis
eds. K.-C. Chang and Y. Long

Vol. 7 Progress in Variational Methods
eds. C. Liu and Y. Long

Vol. 8 Quantized Algebra and Physics
Proceedings of International Workshop
eds. M.-L. Ge, C. Bai and N. Jing

Vol. 9 Operads and Universal Algebra
Proceedings of the International Conference
eds. C. Bai, L. Guo and J.-L. Loday

Vol. 10 Emerging Topics on Differential Equations and Their Applications
Proceedings on Sino-Japan Conference of Young Mathematicians
eds. H. Chen, Y. Long and Y. Nishiura

Vol. 11 Symmetries and Groups in Contemporary Physics
Proceedings of the XXIX International Colloquium on Group-Theoretical Methods in Physics
eds. C. Bai, J.-P. Gazeau and M.-L. Ge

ankai
ries in
ure,
plied
ematics
and
retical
ysics

ol. 5

Proceedings of the Special Program at
Nankai Institute of Mathematics

Tianjin, China **January – June 1991**

COMPUTER MATHEMATICS

Edited by

Wu Wen-Tsün
Institute of Systems Science
Academia Sinica

Hu Guo-Ding
Nankai Institute of Mathematics

World Scientific
Singapore • New Jersey • London • Hong Kong

Published by

World Scientific Publishing Co. Pte. Ltd.
5 Toh Tuck Link, Singapore 596224
USA office: 27 Warren Street, Suite 401-402, Hackensack, NJ 07601
UK office: 57 Shelton Street, Covent Garden, London WC2H 9HE

British Library Cataloguing-in-Publication Data
A catalogue record for this book is available from the British Library.

Nankai Series in Pure, Applied Mathematics and Theoretical Physics — Vol. 5
COMPUTER MATHEMATICS
Proceedings of the Special Program at Nankai Institute of Mathematics

ISBN-13 978-981-02-1528-6
ISBN-10 981-02-1528-2

PREFACE

In each academic year the Nankai Institute of Mathematics will focus on one or two particular topics for its activities. In the present era of computers it is important for mathematics to be developed in a manner to take advantage of the powerful tool — the computer. Due to this reason, in spring 1991 the Nankai Institute has chosen computer mathematics as one of its topics of activities. During this period some regular courses on mathematics mechanization, computation complexity, and computer algebra system were given by Professors H. Shi, W. Q. Huang, and R. B. Zhong respectively. The audience consists of mainly young scholars from various parts of China. Several foreign guests including Professors *T. Y. Li*, *C. Bajaj* and *G. E. Collins* of USA, *E. Engeler* of Switzerland, *V. P. Gerdt* of USSR, *M. Mignotte* of France, have been invited to give lectures of particular interest on computer mathematics. A seminar had also been held in this period at the end of May, bearing the special name of "mathematics mechanization". The present Proceedings include selected papers presented in the seminar as well as lectures given by the invited speakers. We regret that Professors S. C. Chou and J. R. Gabriel were not able to come to China. We hope that the publication of this Proceedings may serve to enchance somewhat the tendency of closer and closer connection between computer and mathematics.

Wu Wen-tsün

CONTENTS

Computer Mathematics
Proc. of the Special Program at
Nankai Institute of Mathematics
January 1991 - June 1991

On the Development of Polynomial Equations Solving in China*

Wu Wen-tsun
Institute of Systems Science, Academia Sinica

1. Introduction.

In Chap. 2 of the classic ≪Mathematical Discovery≫ Polya has called our attention about a posthumous work of Descartes, ≪Rules for the Direction of the Mind≫. Polya described a rough outline of the scheme that Descartes expected to be applicable to all types of problems which we reproduce below:

> First, reduce any kind of problem to a mathematical problem.
>
> Second, reduce any kind of a mathematical problem to a problem of algebra.
>
> Third, reduce any problem of algebra to the solution of a single equation.

The third step may be further clarified as shown in the paragraph below which we again reproduce from the above classic of Polya:

> Descartes advises us to set up as many equations as there are unknowns. Let n stand for the number of unknowns, and $x_1, x_2, \cdots, x_n$ for the unknowns themselves; then we can write the desired system of equations in the form

$$
\begin{aligned}
r_1(x_1, x_2, \cdots, x_n) &= 0,\\
r_2(x_1, x_2, \cdots, x_n) &= 0,\\
&\cdots\cdots,\\
r_n(x_1, x_2, \cdots, x_n) &= 0,
\end{aligned}
$$

*The present paper is partially supported by NSFC Grant JI85312 and TWAS Grant 86-93

> where $r_1(x_1, x_2, \cdots, x_n)$ indicates a polynomial in $x_1, x_2, \cdots, x_n$, and the left-hand sides of the following equations must be similarly interpreted. Descartes advises us further to reduce this system of equations to one final equation.

It seems that Descartes was absolutely absurd in reducing any kind of problem to one of polynomial equations solving. Even the last step of reducing such a system of polynomial equations to one single final equation is not at all evident how to do and is actually entirely non-trivial. However, let us reproduce again some paragraph from the same classic:

> Descartes' project failed, but it was a great project and even in its failure it influenced science much more than a thousand and one little projects which happened to succeed. Although Descartes' scheme does not work in all cases, it does work in an inexhaustible variety of cases, among which there is an inexhaustible variety of *important* cases.

In fact, in now-a-days we see that there are really inexhaustive problems from mathematics, from sciences, from technologies, etc. which will ultimately be reduced to polynomial equations solving. Among all kinds of systems of equations to be solved, those of polynomial forms are clearly the simplest, the most fundamental, and at the same time entirely non-trivial one. Among all kinds of problems in mathematics, the problem of polynomial equations solving should therefore be considered as one of primary importance to be settled with highest priority.

2. Polynomial Equations Solving in Ancient China.

In contrast to the Euclidean system of ancient Greek for which theorem-proving was the main concern, equations-solving occupied the central position in Chinese ancient mathematics. Below we shall give a brief outline of the main achievements of our ancestors on this subject.

The earliest mathematical classic of ancient China which fortunately had been preserved up to present day was ≪ *Nine Chapters in Arithmetic* ≫ (abbr. *NC)*, completed about 1c B.C. In that classic there had already appeared the method of solving a system of linear equations together with the introduction of negative numbers which is actually the same as the present-day Gaussian elimination method. The coefficients of linear equations were represented by well-arranged counting rods put in a matrix form on a counting board. By manipulating the counting rods in some mechanical way the matrix of coefficients will be reduced to a form which in modern notations is in general equivalent to a set of equations as shown below:

$$\begin{aligned}
c_{11} * x_1 + d_1 &= 0,\\
c_{22} * x_2 + c_{21} * x_1 + d_2 &= 0,\\
\dots\\
c_{nn} * x_n + \cdots + c_{n1} * x_1 + d_n &= 0,
\end{aligned} \tag{2.1}$$

In the above equations (eventually some may be lacking as shown by one example in *NC*) all the leading coefficients c_{ii} are non-zero and the solutions are then readily found (eventually in terms of some parameters corresponding to the leading variables x_{ii} of the missing equations).

In another classic ≪ *Annotations to NC* ≫ (abbr. *AN)* appeared in 263A.D. by Liu Hui there were shown some variants of the above elimination method, comparison of efficiency of these variants, and detailed computations of some concrete equation-system in 5 unknowns.

There was already some trace of quadratic equations in *NC*. In fact, the extraction of square root corresponding to the solving of simplest quadratic equation $x^2 = a$ was solved by means of some geometrical principle, viz. the so-called *Out-In Complementary Principle*. This geometrical method of square-root extraction led to the solving of more general quadratic equations. This may be seen from most of our ancient classics that instead of saying as in our modern language *solving quadratic equation of the form* $x^2 + a * x = b$, *one said extracting square root of b with* cong *a*. In ≪ *NC* ≫ there was even a geometrical problem which was reduced to the square-root extraction with *cong*. In ≪ *AN* ≫ Liu Hui had given proofs of the validity of the above methods. In a classic of Zhao Shuang in about the same period as Liu Hui it was implicitely indicated how to express the solutions of a quadratic equation in terms of its coefficients. However, we have to remark that our ancestors seems to have little interest in the general expression of roots of equations in terms of coefficients of the equations. They were rather interested in the determination of their real numerical solutions. This attitude of meeting practical applications permeated the whole period of later development of equations-solving in our ancient time.

The geometrical method of square-root extraction worked well also for cubic-root extraction corresponding to the solving of cubic equations of particular form $x^3 = a$, as already shown in ≪ *NC* ≫ as well as ≪ *AN* ≫. It was natural to extend the method to the solving of general cubic equations, as in the case of quadratic equations. Probably the celebrated mathematician Zu Congzhi of 5c had tried to solve such general cubic equations, though it was not quite certain since no written records were preserved. However, in the beginning of Tang Dynasty in 7c there appeared the classic of Wang Xiaotong in which were exhibited complicate problems arising from geometry and practice. After the description of how the final cubic equations should be formed from such concrete problems, Wang gave directly the numerical solutions in saying simply *by root-extraction*. This shows that before the time of Wang the intellectuals should already know how to solve in numerical values a given cubic equation in numerical coefficients.

The way of further development was quite clear. Thus, according to the late renown historian in Chinese ancient mathematics Professor Yan Dongji, in calendars of Sung Dynasty (10c) there were biquadratic equations to be solved numerically. Unfortunately Professor Yan died prematurely in 1988 and no written manuscripts were left to us. In any way, throughout the whole period of Sung Dynasty (10-13c, very likely beginning from much earlier time) efforts were exerted for the solving of polynomial equations of arbitrary degree. The goal was ultimately reached at latest in 13c. In fact, in 1249 there appeared the classic ***Mathematics in Nine Chapters*** of Qin Jiushao of Sung Dynasty. In that classic algorithm for the determination of numerical solutions of a polynomial equation of arbitrary degree was given which is actually the same as the much later re-discovered Horner's method.

During the period of Sung and Yuan Dynasties there was a discovery of utmost importance in the introduction of the precise notion of an *unknown* under the name of *Heaven's Element*. A so-called *Heaven's Element Method* was developed which permitted to turn easily geometrical entities into the form of polynomials and geometrical relations into polynomial equations. It was known that there were texts about treatment of pairs of polynomial equations in two unknowns *Heaven's* and *Earth's Elements* but unfortunately such texts were lost. Fortunately there remained a classic *Four Elements of Jade Mirrors* of Zhu Szejie in 1303 of Yuan Dynasty which treated polynomial equations up to even four unknowns. The method is one of elimination and may be seen clearly from one of the representative examples in Zhu's classic as shown below.

A geometrical problem was led to a set of three equations in three unknowns which in modern notations may be written in the following form (after removal of the factor x which will be supposed to be non-zero for simplicity):

$$\begin{aligned} x * y * z - x * y^2 - z - x - y &= 0, \\ x * z - x^2 - z - y + x &= 0, \\ z^2 - x^2 - y^2 &= 0. \end{aligned}$$

Zhu's method runs briefly as follows. Eliminate first y to get two equations in x and z alone. Eliminate then x to get a single equation in z alone. Choose from these equations three to be written in the modern form below:

$$\begin{aligned} z^4 - 6 * z^3 + 4 * z^2 + 6 * z - 5 &= 0 \\ x^3 - (z+2) * x^2 + (z^2 + 4 * z + 2) * x - (2 * z^2 + 2 * z) &= 0, \\ y - (x * z - x^2 - z + x) &= 0. \end{aligned} \tag{2.2}$$

Solve the equations (2.2) successively in single unknowns z, x, and y we get one of the solutions to be $(z, x, y) = (5, 3, 4)$. In general Qin Jiushao's method may be applied to get all possible real solutions if required.

3. Polynomial Equations-Solving in MMRC of Modern China.

Let K be a field of characteristic 0 and $x = (x_1, \ldots, x_n)$ be a set of indeterminates to be fixed in what follows. By a *pol* will be meant a polynomial in the ring $K[x_1, \cdots, x_n]$ and by a *polset* a finite set of such pols. For a polset PS consisting of non-zero pols $P_1, \cdots, P_n$ we shall write simply $PS = 0$ for the set of equations $P_1 = 0, \cdots, P_n = 0$. The totality of all possible solutions of $PS = 0$ or zeros of PS in all conceivable extension fields of K will be denoted by $Zero(PS)$. For any other pol G we shall write $Zero(PS/G)$ for the set $Zero(PS) - Zero(G)$, i.e. the totality of zeros of PS which are not zeros of G. The problem of polynomial equations-solving amounts then to the determination of the set $Zero(PS)$ for any given polset PS.

The equations (2.1) and (2.2) serve now as a clue for the solving of polynomial equations which had been carried out in recent years by the Mathematics-Mechanization Research Center (MMRC) of Institute of Systems Science in our Academy, cf. e.g. [WU1,2].

To begin with, let us first introduce the following all the more important concept of an *ascending set* (abbr. *asc-set*) which we have borrowed from the two classics of J.F.Ritt, cf. [R1,2].

For a non-constant pol P let c be the greatest subscript for which x_c appears actually in P. Let d be then the degree of P considered as a pol in x_c. Such a pol can then be written in the *canonical form*

$$P = I * x_c^d + \textit{lower degree terms in } x_c$$

with coefficients themselves pols in the ring $K[x_1, \cdots, x_{c-1}]$. We call c the *class* and d the *class-degree* or simply *degree* of P, to be denoted by $cls(P)$ and $cdeg(P)$ or simply $deg(P)$ resp. The leading coefficient I of P w.r.t. the leading variable x_c is then called the *initial* of the pol P. For a non-zero constant pol the class will be defined to be 0 while the degree may be either defined as -1 or may be left undefined.

Consider now a polset AS of the following particular form. Suppose that the indeterminates $x_1, \cdots, x_n$ can be rearranged as $u_1, \cdots, u_d, y_1, \cdots, y_r$ $(r + d = n)$ such that the pols in AS can be arranged as $A_1, \cdots, A_r$ in canonical forms below:

$$
\begin{aligned}
A_1 &= I_1 * y_1^{d_1} \;+\; lower\ degree\ terms\ in\ y_1, \\
A_2 &= I_2 * y_2^{d_2} \;+\; lower\ degree\ terms\ in\ y_2, \\
&\cdots\cdots \\
A_r &= I_r * y_r^{d_r} \;+\; lower\ degree\ terms\ in\ y_r.
\end{aligned} \tag{3.1}
$$

In (3.1) for each i the coefficients of y_i-powers in A_i are pols in $K[u_1, \cdots, u_d, y_1, \cdots, y_{i-1}]$. We suppose also each pol I_i has a degree in y_j less than $deg(A_j) = d_j$ for any pair (i, j) with $j < i$. In that case we say that the polset AS is an ***ascending set*** and the coefficient pols I_i are *initials* of the corresponding pols A_i. Note that in returning back to original notations, y_i are just x_{c_i} with $c_i = cls(A_i)$ where

$$0 < c_1 < \cdots < c_r,$$

while $u = (u_1, \cdots, u_d)$ are the remaining $x's$ arranged say in the same order as before.

In comparing with the equations (2.1) and (2.2) in Sect 1, we see clearly that an asc-set defined by (3.1) is just a generalization of the polsets occuring in the left-side of (2.1) and (2.2).

A polset consisting of a single non-zero constant will be called a *trivial* asc-set. An asc-set of form (3.1) will then be said to be a *non-trivial* one. It is clear that the zero-set of a trivial asc-set is empty, and for a non-trivial asc-set AS the zero-set $Zero(AS)$ may also be considered as completely determined.

Our method of solving polynomial equations consists of now in reducing the problem of solving arbitrary polynomial equations to those of solving such equations corresponding to asc-sets. One of the variants of such method is expressed in the theorem below.

Main Theorem. There is an algorithm which permits to determine for any polset PS a finite set of asc-sets AS_i in a finite number of steps such that

$$Zero(PS) \;=\; SUM_i\ Zero(AS_i/IP_i), \tag{3.2}$$

in which each IP_i is the product of all initials of pols in the corresponding AS_i.

The formulas (3.2) as well as its variants will be called *decomposition formulas* for $Zero(PS)$. The corresponding Main Theorems and the decomposition formulas form the corner-stones on which are based our method of polynomial equations-solving as well as all the applications. We leave these to the relavant papers published mostly in the irregular publication Mathematics Mechanization Research Preprints (abbr. MM-

Preprints).

REFERENCES.

[R1] J.F. Ritt, Differential equations from the algebraic standpoint, Amer.Math.Soc., Providence (1932).

[R2] J.F. Ritt, Differential algebra, Amer.Math.Soc., Providence (1950).

[WU1] Wu Wen-tsun, On zeros of algebraic equations — an application of Ritt principle, Kexue Tongbao 31 (1986) 1-5.

[WU2] Wu Wen-tsun, A zero structure theorem for polynomial-equations-solving and its applications, MM-Res.Preprints, No.1 (1987) 2-12.

Computer Mathematics
Proc. of the Special Program at
Nankai Institute of Mathematics
January 1991 – June 1991

The Possibility of Using Computer to Study the Equation of Gravitation

Qi-keng Lu
Institute of Mathematics, Academia Sinica
Beijing, China 100080

This paper is a review of our endeavour in the middle of 1970s and a heuristic statement of the possibility in the future to study the equation of gravitation, rather than a conclusion and a detailed proof of results.

Let M be a 4-dimensional differentiable manifold which possesses a Lorentz metric

$$ds^2 = \sum_{j,k=0}^{3} g_{jk} dx^j dx^k, \tag{1}$$

that means the 4×4 matrix

$$G = (g_{jk})_{0 \le j,k \le 3} = A \begin{pmatrix} 1 & & & \\ & -1 & & \\ & & -1 & \\ & & & -1 \end{pmatrix} A^T$$

where A is a non-singular 4×4 matrix and A^T its transposed matrix. Denote

$$\left\{ {i \atop jk} \right\} = \sum_{l=0}^{3} g^{il} \left(\frac{\partial g_{jl}}{\partial x^k} + \frac{\partial g_{kl}}{\partial x^j} - \frac{\partial g_{jk}}{\partial x^l} \right) \tag{2}$$

as the Christoffel symbol or the pseudo-Riemann connection coefficients. The (pseudo-) Riemann curvature tensor is defined by

$$R^i_{jkl} = \frac{\partial}{\partial x^k} \left\{ {i \atop jl} \right\} - \frac{\partial}{\partial x^l} \left\{ {i \atop jk} \right\} + \sum_{r=0}^{3} \left\{ {i \atop rk} \right\} \left\{ {r \atop jl} \right\} - \sum_{r=0}^{3} \left\{ {i \atop rl} \right\} \left\{ {r \atop jk} \right\} \tag{3}$$

and the Ricci tensor is

$$R_{jl} = \sum_{i=0}^{3} R^i_{jil} \tag{4}$$

and $R = \sum_{j,l} g^{jl} R_{jl}$.

The Einstein equation of gravitation is

$$R_{jl} - \frac{1}{2} g_{jl} R + \wedge g_{jl} = T_{jl} \tag{5}$$

where $\wedge$ is the cosmology constant and T_{jl} is the energy-momentum tensor, which satisfies the equation of conservation

$$\sum_{j,k} g^{jk} T_{ij;k} = 0. \tag{6}$$

Note that g^{jk} are the elements of the inverse matrix G^{-1} and the semicolon ";" of a tensor is its covariant differentiation with respect to the Riemann connection. More precisely,

$$T_{ij;k} = \frac{\partial T_{ij}}{\partial x^k} - \sum_{l=0}^{3} T_{lj} \left\{ {l \atop ik} \right\} - \sum_{l=0}^{3} T_{il} \left\{ {l \atop jk} \right\}. \tag{7}$$

In eq. (5), g_{jl}'s are considered as unknown quantities. Though globally in a differential manifold a Lorentz metric need not exist (c.f.[Li]), in most works of general relativity the solutions are local in nature. The energy-momentum tensor T_{jl} is in general considered as "given" quantities in the sense that its expressions are known but the unknown quantities g_{jl} may be contained in T_{jk}, and eq. (6) is in fact the condition of integrability because the left-hand side of (5) satisfies

$$g^{lk}(R_{jl} - \frac{1}{2} g_{jl} R + \wedge g_{jl})_{;k} = 0$$

identically.

Since the Einstein equation is highly non-linear, usually some symmetric or physical condition on g_{jl}'s is assumed and then the Ricci tensor R_{jl} is caculated in order to see what simpler equations can be deduced. However, from definitions (2–4), one can see that it is obviuosly a tedious and laborious task. Naturally, such a mechanical process should be left to the computer, if possible, and spare the brain for the most important problems.

In early 1970s, I read some papers of mathematical physics, in which the authors stated that they used the computer as an auxiliary tool to solve Einstein's equations. But they did not mention what type of computer or what kind of language they used. Starting 1976, I bagan to write programs of formula manipulation by LISP. However, on the outdated computer DSJ-21, the earliest made in China, there was no LISP language available . Mr. Dongyue Chen developed an FCY language and realized it in DSJ-21. We used FCY to establish formal algebraic computation such as polynomial operations (including finding the largest common divisor of polynomials of several variables), rational function operations, rational transformations, differentiation of elementary functions, etc. The result was a joint-paper [LC]. But I should say that the most tedious and laborious work was done by Chen. Besides, he used FCY successfully to solve several practical problems, including proving some hard theorems of elementary geometry by Wu's method. These theorems were failed to prove even by much bigger and faster computers in U.S. Our next step was the manipulation of tensor calculus. But we stopped because, firstly, it needed many man-years while we were only two and our computer was so obsolete and secondly we later knew that there already existed abroad some powerful languages suitable for formula manipulation, but we could not afford to buy such software and a bigger and faster computer. After more than ten years, I saw the success of Wu's method, which is now being extended to solve differential equations of polynomial type. It inspires me to speculate whether it is possible to solve the equation of gravitation in spinor form by Wu's method [W] to a certain extent.

According to my experience, solving the equation of gravitation in spinor form contains a lot of mechanical calculation, which may be done more suitably on a computer. Especially, solving such equations in spinor form is a successive process of solving systems of partial differential equations of first order, which are not linear but appear in a polynomial manner.

Since spinor analysis is not much popular among physicists or even mathematicians, let us outline some of the theory and notations.

Let the tangent bundle of a Lorentz manifold M^4be $T(M)$, the structure group of which is SO(1,3), i.e., the Lorentz group. Then there is an associated vector bundle Spin(M), called the spinor bundle, the structure group of which is SL(2,C). It is well-known that there is a two to one map

$$SL(2,C) \to SO(1,3)$$

which is a local isomorphism. Denote $Spin^*(M)$ as the dual vector bundle of spin(M) and $\overline{Spin}(M)$ the complex conjugate vector bundle. The spinor analysis is the study of tensor bundles

$$[Spin(M)]^p \otimes [\overline{Spin}(M)]^q \otimes [Spin^*(M)]^r \otimes [\overline{Spin}^*(M)]^S$$

where p,q,r,s are non-negative integers. The relation between such bundles and the usual tensor bundles

$$[T(M)]^m \otimes [T^*(M)]^n$$

will be explained in the following.

Let

$$\{ X_a = \textstyle\sum_{j=0}^{3} e^j_{(a)}(x) \frac{\partial}{\partial x^j} \}_{a=0,1,2,3}$$

be a Lorentz frame of M and

$$\{ \omega^a = \textstyle\sum_{j=0}^{3} e^{(a)}_j dx^j \}_{a=0,1,2,3}$$

be its co-frame. That means

$$\omega^a(X_b) = \delta^a_b \text{ and } ds^2 = (\omega^0)^2 - (\omega^1)^2 - (\omega^2)^2 - (\omega^3)^2.$$

Then the coefficients of the pseudo-Riemann connection with respect to the frame X_a are

$$\Gamma^a_{bc} = \sum_j \left[\sum_j \frac{\partial e^j_{(b)}}{\partial x^k} e^{(a)}_j + \sum_{j,l} \left\{ \begin{matrix} l \\ jk \end{matrix} \right\} e^j_{(b)} e^{(a)}_l \right] e^k_{(c)}. \tag{8}$$

Denote the 2×2 hermitian matrices as

$$(\sigma^0_{A\overline{B}}) = \frac{1}{\sqrt{2}} \begin{pmatrix} 1 & 0 \\ 0 & 1 \end{pmatrix}, (\sigma^1_{a\overline{B}}) = \frac{1}{\sqrt{2}} \begin{pmatrix} 0 & 1 \\ 1 & 0 \end{pmatrix},$$

$$(\sigma^2_{A\overline{B}}) = \frac{1}{\sqrt{2}} \begin{pmatrix} 0 & i \\ -i & 0 \end{pmatrix}, (\sigma^3_{A\overline{B}}) = \frac{1}{\sqrt{2}} \begin{pmatrix} 1 & 0 \\ 0 & -1 \end{pmatrix}, \tag{9}$$

which are the Pauli matrices multiplied by a factor $\frac{1}{\sqrt{2}}$. Their complex conjugate matrices are denoted by

$$(\sigma_0^{A\overline{B}}) = \frac{1}{\sqrt{2}} \begin{pmatrix} 1 & 0 \\ 0 & 1 \end{pmatrix}, (\sigma_1^{A\overline{B}}) = \frac{1}{\sqrt{2}} \begin{pmatrix} 0 & 1 \\ 1 & 0 \end{pmatrix},$$

$$(\sigma_2^{A\overline{B}}) = \frac{1}{\sqrt{2}} \begin{pmatrix} 0 & -i \\ i & 0 \end{pmatrix}, (\sigma_3^{A\overline{B}}) = \frac{1}{\sqrt{2}} \begin{pmatrix} 1 & 0 \\ 0 & -1 \end{pmatrix}. \tag{10}$$

Here the indices $A, B, C, D, \ldots$ run from 1 to 2 and $j, k, l, \ldots$ run from 0 to 3.

Then a Lorentz frame $\{X_a\}$ corresponds to a spinor frame $\{X_{A\overline{B}}\}$ such that

$$X_{A\overline{B}} = \sum_{a=0}^{3} \sigma^a_{A\overline{B}} X_a \tag{11}$$

and the corresponding co-frame is

$$\omega^{A\overline{B}} = \sum_{a=0}^{3} e^{A\overline{B}}_{(a)} \omega^a. \tag{12}$$

Two Lorentz frames $\{X_a\}$ and $\{X'_a\}$ in the same coordinate neighbourhood must be related by a Lorentz transformation

$$X'_a = \sum_{b=0}^{3} l_a^b X_b \qquad (l_a^b) \in SO(1,3) \tag{13}$$

and correspondingly, there is at least one matrix $(\alpha_B^A) \in SL(2,C)$ such that

$$X'_{A\overline{B}} = \sum_{C,D=1}^{2} \alpha_A^C \overline{\alpha_B^D} X_{C\overline{D}}. \tag{14}$$

Notice that, when α_A^B is changed to $-\alpha_A^B$, the above relation remains the same.

The coefficients of the pseudo-Riemann connection correspond to the spin coefficients

$$\Gamma_{ABC\overline{D}} = \frac{1}{2} \sum_{E,F=1}^{2} \sum_{a,b,c=0}^{3} \epsilon_{AE} \Gamma_{bc}^a \sigma_a^{E\overline{F}} \sigma_{B\overline{F}}^b \sigma_{C\overline{D}}^c \tag{15}$$

where $\epsilon_{AB} = -\epsilon_{BA} = \epsilon^{AB}$, $\epsilon_{12} = 1$ and $\sigma_a^{A\overline{B}} = \sum_j \sigma_j^{A\overline{B}} e_{(a)}^j$, $\sigma_{B\overline{F}}^a = \sum_j \sigma_{A\overline{B}}^j e_j^{(a)}$.

Newman and Penrose [NP], first wrote the Einstein equation (5) for $\Lambda = 0$ in spinor form. First, it is easy to see from (15) that

$$\Gamma_{ABC\overline{D}} = \Gamma_{BAC\overline{D}}. \tag{16}$$

Then there are at most twelve independent complex quantities as in the following list:

$\Gamma_{ABC\overline{D}} =$ (17)

	AB		
$C\overline{D}$	11	12 or 21	22
$1\overline{1}$	κ	ε	π
$2\overline{1}$	ρ	α	λ
$1\overline{2}$	σ	β	μ
$2\overline{2}$	τ	γ	ν

Denote

$$D = X_{1\overline{1}}, \Delta = X_{2\overline{2}}, \delta = X_{1\overline{2}}, \overline{\delta} = X_{2\overline{1}}. \tag{18}$$

Then the Einstein equation (5) can be reduced to a system of differential equations of spin coefficients $\Gamma_{ABC\overline{D}}$, which was called the NP equation by later authors. Substituting the Einstein equation (5) into the Weyl tensor

$$C_{abcd} = R_{abcd} - \frac{1}{2}(g_{ac}R_{db} - g_{ad}R_{cb} - g_{bc}R_{da} + g_{bd}R_{ca}) + \frac{R}{6}(g_{ac}g_{db} - g_{ad}g_{cb}) \tag{19}$$

yields (for $\Lambda = 0$)

$$C_{abcd} = R_{abcd} - \frac{1}{2}(g_{ac}T_{bd} - g_{cb}T_{ad} - g_{ad}T_{bc} + g_{bd}T_{ac}) + \frac{R}{3}(g_{ac}g_{bd} - g_{cb}g_{ad}). \tag{20}$$

The spinor form of the Weyl tensor and the Riemann curvature tensor can be expressed in the following manner (c.f.[Lu]) respectively:

$$C_{A\overline{E}B\overline{F}C\overline{G}D\overline{H}} = C_{abcd}\sigma^a_{\ A\overline{E}}\sigma^b_{\ B\overline{F}}\sigma^c_{\ C\overline{G}}\sigma^d_{\ D\overline{H}} = \Psi_{ABCD}\epsilon_{EF}\epsilon_{GH} + \overline{\Psi}_{EFGH}\epsilon_{AB}\epsilon_{CD} \tag{21}$$

where

$$\Psi_{ABCD} = \frac{1}{4}\sum_{E,F,G,H} C_{A\overline{E}B\overline{F}C\overline{G}D\overline{H}}\epsilon^{EF}\epsilon^{GH} \tag{22}$$

is symmetric with respect to all indices A, B, C, D;

$$\begin{aligned} R_{A\overline{E}B\overline{F}C\overline{G}D\overline{H}} &= R_{abcd}\sigma^a_{\ A\overline{E}}\sigma^b_{\ B\overline{F}}\sigma^c_{\ C\overline{G}}\sigma^d_{\ D\overline{H}} \\ &= \chi_{ABCD}\epsilon_{EF}\epsilon_{GH} + \epsilon_{AB}\epsilon_{CD}\overline{\chi}_{EFGH} + \varphi_{AB\overline{G}\overline{H}}\epsilon_{CD}\epsilon_{EF} + \epsilon_{AB}\epsilon_{GH}\overline{\varphi}_{EF\overline{C}\overline{D}} \end{aligned} \tag{23}$$

where

$$\begin{aligned} \chi_{ABCD} &= \frac{1}{4}R_{A\overline{E}B\overline{F}C\overline{G}D\overline{H}}\epsilon^{EF}\epsilon^{GH} \\ \varphi_{AB\overline{G}\overline{H}} &= \frac{1}{4}R_{A\overline{E}B\overline{F}C\overline{G}D\overline{H}}\epsilon^{EF}\epsilon^{CD} \end{aligned} \tag{24}$$

satisfy

$$\begin{aligned} \chi_{ABCD} &= \chi_{BACD} = \chi_{ABDC} = \chi_{CDBA}, \\ \varphi_{AB\overline{E}\overline{F}} &= \varphi_{BA\overline{E}\overline{F}} = \varphi_{AB\overline{F}\overline{E}} = \overline{\chi_{EF\overline{A}\overline{B}}}. \end{aligned} \tag{25}$$

The spin form of the energy-momentum tensor is

$$T_{A\overline{C}B\overline{D}} = T_{ab}\sigma^a_{\ A\overline{C}}\sigma^b_{\ B\overline{D}}. \tag{26}$$

Changing equation (20) into spinor form and substituting (23), (24) and (26) into it, we have

$$\Psi_{ABCD} = \chi_{ABCD} + \frac{T}{24}(\epsilon_{AD}\epsilon_{BC} + \epsilon_{AC}\epsilon_{BD}) \tag{27}$$

where $T = T_{A\overline{C}B\overline{D}}\epsilon^{AB}\epsilon^{CD}$ and

$$\varphi_{AB\overline{C}\overline{D}} = -\frac{1}{4}(T_{A\overline{C}B\overline{D}} + T_{B\overline{C}A\overline{D}}). \tag{28}$$

$\Gamma_{ABC\overline{D}}$ are the coefficients of a connection on the vector bundle Spin(M). Then the curvature of this connection can be defined:

$$\begin{aligned} F_{ABC\overline{E}D\overline{F}} &= -X_{D\overline{F}}\Gamma_{ABC\overline{E}} + X_{C\overline{E}}\Gamma_{ABD\overline{F}} \\ &+ \Gamma_{AGC\overline{E}}\epsilon^{HG}\Gamma_{HBD\overline{F}} - \Gamma_{AGD\overline{F}}\epsilon^{HG}\Gamma_{HBC\overline{F}} + \Gamma_{ABG\overline{E}}\epsilon^{HG}\Gamma_{HCD\overline{F}} \\ &- \Gamma_{ABG\overline{F}}\epsilon^{HG}\Gamma_{HDC\overline{E}} + \Gamma_{ABC\overline{H}}\epsilon^{GH}\overline{\Gamma_{GEF\overline{D}}} - \Gamma_{ABD\overline{H}}\epsilon^{GH}\overline{\Gamma_{GFE\overline{C}}}. \end{aligned} \tag{29}$$

On the other hand, it can be proved([Lu] Th. 1.3.4) that

$$F_{ABC\overline{E}D\overline{F}} = \chi_{ABCD}\epsilon_{\overline{EF}} + \varphi_{AB\overline{EF}}\epsilon_{CD}. \tag{30}$$

Substituting (30) into the left-hand side of (29) and expressing χ_{ABCD} by Ψ_{ABCD} according to (27) yield the NP equations when the indices A,B,....,F run from 1 to 2 and the notation of (17) is used.

NP Equations

$$D\rho - \overline{\delta}\kappa = \rho^2 + \sigma\overline{\sigma} + (\epsilon + \overline{\epsilon})\rho - \overline{\kappa}\tau - \kappa(3\alpha + \overline{\beta} - \pi) + \varphi_{11\overline{11}}$$

$$D\sigma - \delta\kappa = (\rho + \overline{\rho})\sigma + (3\epsilon - \overline{\epsilon})\sigma - (\tau - \overline{\pi} + 3\beta + \overline{\alpha})\kappa + \Psi_{1111}$$

$$D\tau - \Delta\kappa = (\tau + \overline{\pi})\rho + (\overline{\tau} + \pi)\sigma + (\epsilon - \overline{\epsilon})\tau - (3\gamma + \overline{\gamma})\kappa + \Psi_{1112} + \varphi_{11\overline{12}}(*)$$

$$D\alpha - \overline{\delta}\epsilon = (\rho + \overline{\epsilon} - 2\epsilon)\alpha + \beta\overline{\sigma} - \overline{\beta}\epsilon - \overline{\kappa}\gamma - \kappa\lambda + (\epsilon + \rho)\pi + \varphi_{12\overline{11}}$$

$$D\beta - \delta\epsilon = (\alpha + \pi)\sigma + (\overline{\rho} - \overline{\epsilon})\beta - (\mu + \gamma)\kappa - (\overline{\alpha} - \overline{\pi})\epsilon + \Psi_{1112}$$

$$D\gamma - \Delta\epsilon = (\tau + \overline{\pi})\alpha + (\overline{\tau} + \pi)\beta - (\epsilon + \overline{\epsilon})\gamma - (\gamma + \overline{\gamma}) + \tau\pi - \nu\kappa + \Psi_{1122} + \varphi_{12\overline{12}} + \tfrac{1}{24}T$$

$$D\lambda - \overline{\delta}\pi = \rho\lambda + \overline{\sigma}\mu + \pi^2 + (\alpha - \overline{\beta})\pi - \nu\overline{\kappa} - (3\epsilon - \overline{\epsilon})\lambda + \varphi_{22\overline{12}}$$

$$D\mu - \delta\pi = \overline{\rho}\mu + \sigma\lambda + \pi\overline{\pi} - (\epsilon + \overline{\epsilon})\mu - \pi(\overline{\alpha} - \beta) - \nu\kappa + \Psi_{1122} - \tfrac{1}{12}T$$

$$D\nu - \Delta\pi = (\pi + \overline{\tau})\mu + (\overline{\pi} + \tau)\lambda + (\gamma - \overline{\gamma})\pi - (3\epsilon + \epsilon)\nu + \Psi_{1222} + \varphi_{22\overline{12}}(*)$$

$$\Delta\lambda - \delta\nu = -(\mu + \overline{\mu})\lambda - (3\gamma - \overline{\gamma})\lambda + (3\alpha + \overline{\beta} + \pi - \overline{\tau})\nu - \Psi_{2222}$$

$$\delta\rho - \overline{\delta}\sigma = \rho(\overline{\alpha} + \beta) - \sigma(3\alpha - \overline{\beta}) + (\rho - \overline{\rho})\tau + (\mu - \overline{\mu})\kappa - \Psi_{1112} + \varphi_{111\overline{2}}(*)$$

$$\delta\alpha - \overline{\delta}\beta = \mu\rho - \lambda\sigma + \alpha\overline{\alpha} + \beta\overline{\beta} - 2\alpha\beta + \gamma(\rho - \overline{\rho}) + \epsilon(\mu - \overline{\mu}) - \Psi_{1122} - \tfrac{T}{24} + \varphi_{121\overline{2}}$$

$$\delta\lambda - \overline{\delta}\mu = (\rho - \overline{\rho})\nu + (\mu - \overline{\mu})\pi + \mu(\alpha + \overline{\beta}) + \lambda(\overline{\alpha} - 3\beta) - \Psi_{1222} + \varphi_{22\overline{12}}(*)$$

$$\delta\nu - \Delta\mu = \mu^2 + \lambda\overline{\lambda} + (\gamma + \overline{\gamma})\mu - \overline{\nu}\pi + (\tau - 3\beta - \overline{\alpha})\nu + \varphi_{22\overline{22}}$$

$$\delta\gamma - \Delta\beta = (\tau - \overline{\alpha} - \beta)\gamma + \mu\tau - \sigma\nu - \epsilon\overline{\nu} - \beta(\gamma - \overline{\gamma} - \mu) + \alpha\overline{\lambda} + \varphi_{12\overline{22}}$$

$$\delta\tau - \Delta\sigma = \mu\sigma + \overline{\lambda}\rho + (\tau + \beta - \overline{\alpha})\tau - (3\gamma - \overline{\gamma})\sigma - \kappa\overline{\nu} + \varphi_{11\overline{22}}(*)$$

$$\Delta\rho - \overline{\delta}\tau = -(\rho\overline{\mu} + \sigma\lambda) + (\overline{\beta} - \alpha - \overline{\tau})\tau + (\gamma + \overline{\gamma})\rho + \nu\kappa - \Psi_{1122} + \tfrac{T}{12}(*)$$

$$\Delta\alpha - \overline{\delta}\gamma = (\rho + \epsilon)\nu - (\tau + \beta)\lambda + (\overline{\gamma} - \overline{\mu})\alpha + (\overline{\beta} - \overline{\tau})\gamma - \Psi_{1222}.$$

There are 12 independent quantities $\Gamma_{ABC\overline{D}}$ but 18 NP equations . It is proved that the equations marked by (*) depend upon other equations [J].

According to (21), the spinor form of Weyl tensor is determined by the quantity Ψ_{ABCD}, which is called the Weyl spinor for convenience and plays an important role in the Petrov classification [NP] of Weyl tensors. In fact, let $Z^A(A = 1, 2)$ be a spinor and set $t = Z^1/Z^2$. Then we have the quartic form

$$\begin{aligned}\varphi &= \sum_{A,B,C,D=1}^{2} \Psi_{ABCD} Z^A Z^B Z^C Z^D \\ &= (Z^2)^4(\Psi_{1111}t^4 + 4\Psi_{1112}t^3 + 6\Psi_{1122}t^2 + 4\Psi_{1222}t + \Psi_{2222} \\ &= \lambda(t + t_1)(t + t_2)(t + t_3)(t + t_4)\end{aligned}$$

where t_1, t_2, t_3, t_4 are the four roots of the polynomial of t. The Petrov classification is as follows

Type I, when the roots are different from each other;

Type II, three roots are different, one of which is a double root;

Type III, two different roots, one of which is a triple root;

Type D, two different roots, each of which is a double root;

Type N, only one quadruple root;

Type O, $\Psi_{ABCD} = 0$.

However the components Ψ_{ABCD} of the Weyl spinor in the NP equations are unknown or partially unknown. To find the equation that the Weyl spinor satisfies, the Bianchi identity can be used

$$R^a_{bcd;e} + R^a_{bde;c} + R^a_{bec;d} = 0,$$

the spinor form of which is [NP]

$$\sum_{D,E} \epsilon^{DE} \chi_{ABCD;E\overline{F}} = \sum_{D,E} \epsilon^{DE} \varphi_{AB\overline{F}\overline{D};C\overline{E}}. \tag{31}$$

According to (27), the above equations can be written as

$$\sum_{D,E} \epsilon^{DE} \Psi_{ABCD;E\overline{F}} - \frac{1}{24}(\epsilon_{BC}T_{;A\overline{F}} + \epsilon_{AC}T_{;B\overline{F}}) = \sum_{D,E} \epsilon^{DE} \varphi_{AB\overline{F}\overline{D};C\overline{E}}. \tag{32}$$

By the notation in (17) and (18), (32) is equivalent to

$D\Psi_{1111} - \overline{\delta}\Psi_{1111} + (4\alpha - \pi)\Psi_{1111} - (2\epsilon + 4\rho)\Psi_{1112} + 3\kappa\Psi_{1122} = D\varphi_{11\overline{1}\overline{1}} - \delta\varphi_{11\overline{1}\overline{1}} + (2\overline{\alpha} + 2\beta - \overline{\pi})\varphi_{11\overline{1}\overline{1}} - 2(\epsilon + \overline{\rho})\varphi_{11\overline{1}\overline{2}} + \overline{\kappa}\varphi_{11\overline{2}\overline{2}} - 2\sigma\varphi_{12\overline{1}\overline{1}} + 2\kappa\varphi_{12\overline{1}\overline{2}}$;

$D\Psi_{1122} - \overline{\delta}\Psi_{1112} - \frac{1}{12}DT + \lambda\Psi_{1111} + 2(\alpha - \pi)\Psi_{1112} - 3\rho\Psi_{1122} + 2\kappa\Psi_{1222} = \overline{\delta}\varphi_{11\overline{1}\overline{2}} - \Delta\varphi_{11\overline{1}\overline{1}} + (2\gamma + 2\overline{\gamma} - \overline{\mu})\varphi_{11\overline{1}\overline{1}} - 2(\alpha + \overline{\tau})\varphi_{11\overline{1}\overline{2}} + \overline{\sigma}\varphi_{11\overline{2}\overline{2}} - 2\tau\varphi_{12\overline{1}\overline{2}} + 2\rho\varphi_{12\overline{1}\overline{2}}$;

$D\Psi_{1222} - \bar{\delta}\Psi_{1122} - \frac{1}{24}\bar{\delta}T + 2\lambda\Psi_{1112} - 3\pi\Psi_{1122} + 2(\epsilon - \rho)\Psi_{1222} + \kappa\Psi_{2222} = \bar{\delta}\varphi_{12\bar{1}\bar{2}} - \Delta\varphi_{12\bar{1}\bar{1}} + \nu\varphi_{11\bar{1}\bar{1}} - \lambda\varphi_{11\bar{1}\bar{2}} + (2\bar{\gamma} - \bar{\mu})\varphi_{12\bar{1}\bar{1}} - 2\bar{\tau}\varphi_{12\bar{1}\bar{2}} + \bar{\sigma}\varphi_{12\bar{2}\bar{2}} - \tau\varphi_{22\bar{1}\bar{1}} + \rho\varphi_{22\bar{1}\bar{2}}$;

$D\Psi_{2222} - \bar{\delta}\Psi_{1222} + 3\lambda\Psi_{1122} - 2(\alpha + 2\pi)\Psi_{1222} + (4\epsilon - \rho)\Psi_{2222} = \bar{\delta}\varphi_{22\bar{1}\bar{2}} - \Delta\varphi_{22\bar{1}\bar{1}} + 2\nu\varphi_{12\bar{1}\bar{1}} - 2\lambda\varphi_{12\bar{1}\bar{2}} + (2\bar{\gamma} - 2\gamma - \bar{\mu})\varphi_{22\bar{1}\bar{1}} + 2(\alpha - \tau)\varphi_{22\bar{1}\bar{2}} + \bar{\sigma}\varphi_{22\bar{2}\bar{2}}$;

$\delta\Psi_{1112} - \Delta\Psi_{1111} + (4\gamma - \mu)\Psi_{1111} - 2(\beta + 2\pi)\Psi_{1112} + 2\sigma\Psi_{1122} = D\varphi_{11\bar{2}\bar{2}} - \delta\varphi_{11\bar{2}\bar{1}} + \bar{\lambda}\varphi_{11\bar{1}\bar{1}} + 2(\beta - \bar{\pi})\varphi_{11\bar{1}\bar{2}} + (2\bar{\epsilon} - 2\epsilon - \bar{\rho})\varphi_{11\bar{2}\bar{2}} - 2\sigma\varphi_{12\bar{1}\bar{2}} + 2\kappa\varphi_{12\bar{2}\bar{2}}$;

$\delta\Psi_{1122} - \Delta\Psi_{1112} - \frac{1}{12}\delta T + \nu\Psi_{1111} + 2(\gamma - \mu)\Psi_{1112} - 3\tau\Psi_{1122} + 2\sigma\Psi_{1222} = \bar{\delta}\varphi_{11\bar{2}\bar{2}} - \Delta\varphi_{11\bar{1}\bar{2}} + \bar{\nu}\varphi_{11\bar{1}\bar{1}} + 2(\gamma - \bar{\mu})\varphi_{11\bar{1}\bar{2}} - (2\alpha - 2\bar{\beta} + \bar{\tau})\varphi_{11\bar{2}\bar{2}} - 2\tau\varphi_{12\bar{1}\bar{2}} + 2\rho\varphi_{12\bar{2}\bar{2}}$;

$\delta\Psi_{1222} - \Delta\Psi_{1122} - \frac{1}{24}\Delta T + 2\nu\Psi_{1112} - 3\mu\Psi_{1122} + 2(\beta - \tau)\Psi_{1222} + \sigma\Psi_{2222} = \bar{\delta}\varphi_{12\bar{2}\bar{2}} - \Delta\varphi_{12\bar{1}\bar{2}} + \nu\varphi_{11\bar{1}\bar{2}} - \lambda\varphi_{11\bar{2}\bar{2}} + \bar{\nu}\varphi_{12\bar{1}\bar{1}} - 2\bar{\mu}\varphi_{12\bar{1}\bar{2}} + (2\bar{\beta} - \bar{\tau})\varphi_{12\bar{2}\bar{2}} - \tau\varphi_{22\bar{1}\bar{2}} + \rho\varphi_{22\bar{2}\bar{2}}$;

$\delta\Psi_{2222} - \Delta\Psi_{1222} + 3\nu\Psi_{1122} - 2(\gamma + 2\mu)\Psi_{1222} + (4\beta - \tau)\Psi_{2222} = \bar{\delta}\varphi_{22\bar{2}\bar{2}} - \Delta\varphi_{22\bar{1}\bar{2}} + 2\nu\varphi_{12\bar{1}\bar{2}} - 3\lambda\varphi_{12\bar{2}\bar{2}} + \bar{\nu}\varphi_{22\bar{1}\bar{1}} - 2(\gamma + \bar{\mu})\varphi_{22\bar{1}\bar{2}} + (2\alpha + 2\bar{\beta} - \bar{\tau})\varphi_{22\bar{2}\bar{2}}$.

The solutions of the NP equations are the spin coefficients, not the metric tensor g_{jk}. In order to obtain g_{jk} from the spin coefficients $\Gamma_{ABC\bar{D}}$, it is sufficient to obtain the Lorentz frame $e^j_{(a)}$ from $\Gamma_{ABC\bar{D}}$. This process is also a process of solving a system of partial differential equations of first order.

$$
\begin{aligned}
\Delta e^k_{(0)} - De^k_{(1)} &= (\gamma + \bar{\gamma})e^k_{(0)} + (\epsilon + \bar{\epsilon})e^k_{(1)} - (\tau + \bar{\pi})\overline{e^k_{(2)}} - (\bar{\tau} + \pi)e^k_{(2)} \\
\delta e^k_{(0)} - De^k_{(2)} &= (\bar{\alpha} + \beta - \bar{\pi})e^k_{(0)} + \kappa e^k_{(1)} - (\bar{\rho} + \epsilon - \bar{\epsilon})e^k_{(2)} - \sigma\overline{e^k_{(2)}} \\
\delta e^k_{(1)} - \Delta e^k_{(2)} &= -\bar{\nu}e^k_{(0)} + (\tau - \bar{\alpha} - \beta)e^k_{(1)} + (\mu - \gamma + \bar{\gamma})e^k_{(2)} + \bar{\lambda}\overline{e^k_{(2)}} \\
\bar{\delta}e^k_{(2)} - \delta\overline{e^k_{(2)}} &= -(\mu - \bar{\mu})e^k_{(0)} - (\rho - \bar{\rho})e^k_{(1)} + (\alpha - \bar{\beta})e^k_{(2)} - (\bar{\alpha} - \beta)\overline{e^k_{(2)}}.
\end{aligned}
$$

More precisely, the frame $e^j_{(a)}$ in the above equation is a linear transformation of a Lorentz frame such that

$$
g^{jk} = e^j_{(0)}e^k_{(1)} + e^j_{(1)}e^k_{(0)} - \overline{e^j_{(2)}}e^k_{(2)} - e^j_{(2)}\overline{e^k_{(2)}}.
$$

At the first sight, the spinor formulation of the equation of gravitation is much more complicated than the usual tensor formulation. But in fact it suggests an efficient tool to solve the equations under some algebraic constraints. In 1969, Kinnersley [K] used NP equations to get all the exact solutions of vacuum gravitational field of type D. These solutions contain many famous exact solutions and even solutions unknown before. This for the first time showed the power of the NP equation. Later we solved the NP equations of type N with an electro-magnatic energy-momentum tensor [CLKL] and the corresponding NP equations of Brans-Dick theory of type III, N and O [LLCK]. We noted that there were many mechanical processes in the calculation for solving such equations.

References

[CLKL] Chen-lon Chou, Yu-fon Liu, Han-ying Kuo and K.H. Look(Qi-keng Lu), Null electromagnetic field and pure gravitational radiation. Reports of Beijing Observatory, 2 (1973), 33 (in Chinese).

[J] Shen Jiang, Simplification of the spinor formulism and its application to Yang-Mills field. Acta Physica, 26(1977), 256 (in Chinese).

[K] W.Kinnersley, Type D "vacuum metrics", J.Math.Phys, 10 (1969) , 1175.

[LC] Qi-keng Lu and Dongyue Chen, Formula manipulation by computer. Acta Computer, 1 (1980), 193. (in Chineses).

[Li] A.Lichnerowicz. Theorie Globale des connexions et des Groupes d'Holonomie. Edizion Cremonese, 1955.

[LLCK] Kai Hung Look(Qi-keng Lu), Yu-fon Liu, Cheng-lon Chou and Han-ying Kuo, The scalar- tensor gravitational wave, 23 (1974) , 95 (in Chinese).

[Lu] Qi-keng Lu, Differential Geometry and Its Application to Physics. Science Press, 1982 (in Chinese).

[NP] T.Newman and R.Penrose, An approach to gravitational radiation by a method of spin coefficients. J.Math.Physics, 3 (1962), 566.

[P] R.Penrose, A spinor approach to general relativity. Ann Phys. (N.Y.), 10 (1960), 171.

[Wu] Wu Wen-tsün, On the foundation of algebraic differential Geometry. Systems Science and Mathemetical Sciences. Vol.2, No.4 (1989), 289.

Computer Mathematics
Proc. of the Special Program at
Nankai Institute of Mathematics
January 1991 – June 1991

Solving polynomial systems by homotopy continuation methods[1]

T. Y. Li
Department of Mathematics
Michigan State University
East Lansing, MI 48824-1027, U.S.A.

1. Introduction

Let $P(x) = 0$ be a system of n polynomial equations in n unknowns. Denoting $P = (p_1, \ldots, p_n)$, we want to find all isolated solutions of

$$\begin{aligned} p_1(x_1, \ldots, x_n) &= 0 \\ &\vdots \\ p_n(x_1, \ldots, x_n) &= 0 \end{aligned} \tag{1}$$

for $x = (x_1, \ldots, x_n)$. In 1977, Garcia and Zangwill [6] and Drexler [5] independently presented theorems suggesting that homotopy continuation could be used to find numerically the full set of isolated solutions of (1). The homotopy continuation method for solving this system is to define a trivial system $Q(x) = (q_1(x), \ldots, q_n(x)) = 0$ and then follow the curves in the real variable t which make up the solution set of

$$0 = H(x, t) = (1 - t)Q(x) + tP(x). \tag{2}$$

More precisely, if $Q(X) = 0$ is chosen correctly, the following three properties hold:

Property 0. (*Triviality*) The solutions of $Q(x) = 0$ are known.

Property 1. (*Smoothness*) The solution set of $H(x, t) = 0$ for $0 \leq t < 1$ consists of a finite number of smooth paths, each parameterized by t in $[0, 1)$.

[1]Research was supported in part by NSF under Grant CCR-9024940.

Property 2. (*Accessibility*) Every isolated solution of $H(x,1) = P(x) = 0$ is reached by some path originating at $t = 0$. It follows that this path starts at a solution of $H(x,0) = Q(x) = 0$.

When the three properties hold, the solution paths can be followed from the initial points (known because of Property 0) at $t = 0$ to all solutions of the original problem $P(x) = 0$ at $t = 1$ using standard numerical techniques. (Cf. [1])

Several authors have suggested the choices of Q that satisfy the three properties. (Cf.[4,5,6,8,24,25] for a partial list.) A typical suggestion is

$$q_1(x) = a_1 x_1^{d_1} - b_1, \quad \cdots, \quad q_n(x) = a_n x_n^{d_n} - b_n \tag{3}$$

where $d_1, \ldots, d_n$ are the degrees of $p_1(x), \ldots, p_n(x)$, and a_i, b_i are random complex numbers. So in one sense, the original problem we posed is solved. All solutions of $P(x) = 0$ are found at the end of the $d_1 \ldots d_n$ paths that make up the solution set of $H(x,t) = 0, 0 \leq t < 1$.

In this article, we report some recent development which make this method more convenient to apply.

The reason the problem is not satisfactorily solved by the above considerations is the existence of extraneous paths. Although the above method produces $d = d_1 \ldots d_n$ paths, the system $P(x) = 0$ may have fewer than d solutions. We call such a system *deficient*. In this case, some of the paths produced by the above method will be extraneous paths.

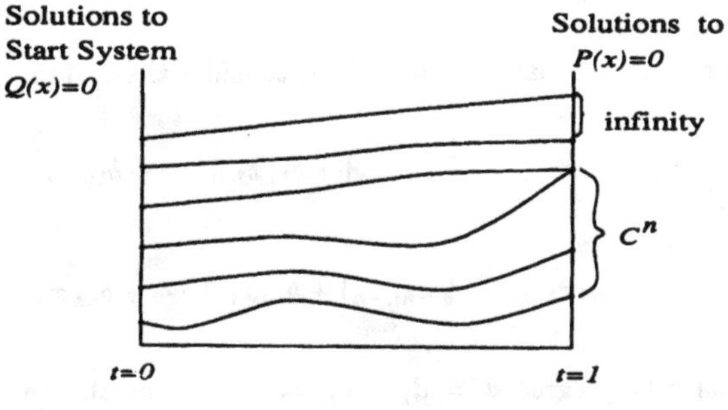

Figure 1.

More precisely, even though properties 0-2 imply that each solution of $P(x) = 0$ will lie at the end of a solution path, it is also consistent with these properties that some of the paths may diverge to infinity as the parameter t approaches 1. (The smoothness property rules this out for $t \to t_0 < 1$.) Otherwise said, it is quite possible for $Q(x) = 0$

to have more solutions than $P(x) = 0$. In this case, some of the paths leading from roots of $Q(x) = 0$ are extraneous, and typically diverge to infinity. (See Figure 1.)

To organize our discussion, we will at times use a notation that makes the coefficients and variables in $P(x) = 0$ explicit. Thus when the dependence on coefficients is important, we will consider the system $P(c, x) = 0$ of n polynomial equations in n unknowns,where $c = (c_1, \cdots, c_M)$ are coefficients and $x = (x_1, \cdots, x_n)$ are unknowns. Two different problems can be posed:

Problem A. For a given choice of coefficients c, solve the system of equations $P(c, x) = 0$.

Problem B. For each of several different choices of coefficients c, solve the system of equations $P(c, x) = 0$.

We divide our discussion on dealing with and eliminating extraneous paths for Problem A in Section 2 and for Problem B in Section3. Some numerical considerations, the use of projective coordinates and real homotopies, are given in Section 4.

2. Methods for Problem A

The progress on Problem A is the least satisfactory among the areas we discuss. For deficient systems, there are some partial results that use algebraic geometry to reduce the number of extraneous paths with varying degree of success.

2.1 Random Product Homotopy

For a specific example that is quite simple, consider the system

$$\begin{aligned} p_1(x) &= x_1(a_{11}x_1 + \cdots + a_{1n}x_n) + b_{11}x_1 + \cdots + b_{1n}x_n + c_1 = 0 \\ &\vdots \\ p_n(x) &= x_1(a_{n1}x_1 + \cdots + a_{nn}x_n) + b_{n1}x_1 + \cdots + b_{nn}x_n + c_n = 0 \end{aligned} \tag{4}$$

This system has total degree $d = d_1 \cdots d_n = 2^n$. Thus the "expected number of solutions" is 2^n, and the classical homotopy continuation method using the start system $Q(x) = 0$ in (3) sends out 2^n paths from 2^ntrivial starting points. However, the system $P(x) = 0$ has only $n+1$ isolated solutions (even fewer for special choices of coefficients). This is a deficient system; at least $2^n - n - 1$ paths will be extraneous. It is never known from the start which of the paths will end up to be extraneous, so that must all be followed to the end, representing wasted computation.

The random product homotopy was developed in [9,10] to alleviate this problem.

According to that technique, a more efficient choice for the trivial system $Q(x) = 0$ is

$$\begin{aligned} q_1(x) =&(x_1 + e_{11})(x_1 + x_2 + \cdots + x_n + e_{12}) \\ q_2(x) =&(x_1 + e_{21})(x_2 + e_{22}) \\ &\vdots \\ q_n(x) =&(x_1 + e_{n1})(x_n + e_{n2}) \end{aligned} \tag{5}$$

Set

$$H(x,t) = (1-t)cQ(x) + tP(x).$$

It is clear by inspection that for a generic choice of the complex numbers e_{ij}, $Q(x) = 0$ has exactly $n+1$ roots. Thus there are only $n+1$ paths starting from $n+1$ starting points for this choice for homotopy . It is proved in [10] that Property 0-2 hold for this choice of $H(x,t)$, for almost all complex numbers e_{ij} and c. Thus all solutions of $P(x) = 0$ are found at the end of $n+1$ paths. The result of [10] is then a mathematical result, that there can be at most $n+1$ solutions to (4), as well as the basis of a numerical procedure for approximating the solutions.

The reason this works is quite simple. The solution paths of (2) which do not proceed to a solution of $P(x) = 0$ in C^n diverge to infinity. If the system (2) is viewed in projective space

$$\mathbf{P}^n = \{(x_0, \cdots, x_n) \in C^{n+1} \setminus (0, \cdots, 0)\}/\sim$$

where the equivalent relation "$\sim$ " is given by $x \sim y$ if $x = cy$ for some nonzero $c \in C$, the diverging paths are simply proceeding to a "point at infinity" in $\mathbf{P}^n$.

For a polynomial $f(x_1, \cdots, x_n)$ of degree d, denote the associated homogeneous polynomial

$$\overline{f}(x_0, x_1, \cdots, x_n) = x_0^d f(\frac{x_1}{x_0}, \cdots, \frac{x_n}{x_0}).$$

The solutions of $f(x) = 0$ at infinity are those zeros of $\overline{f}$ with $x_0 = 0$ and the remaining zeros of $\overline{f}$ with $x_0 \neq 0$ are the solutions of $f(x) = 0$ in C^n.

Viewed in projective space $\mathbf{P}^n$ the system $P(x) = 0$ in (4) has some roots at infinity. The roots at infinity make up a nonsingular variety, specifically the linear space P^{n-2} defined by $x_0 = x_1 = 0$. A Chern class formula from intersection theory shows that the contribution of a linear variety of solutions of dimension e to the "total degree" $(d_1 \times \cdots \times d_n)$, or the total expected number of solutions, of the system is at least s, where s is the coefficient of t^e in the Maclaurin series expansion of

$$(1+t)^{e-n} \prod_{i=1}^{n} (1 + d_i t).$$

In our case, $d_1 = \cdots = d_n = 2$, and $e = n - 2$, hence,

$$\frac{(1+2t)^n}{(1+t)^2} = \frac{(1+t+t)^2}{(1+t)^2} = \frac{\sigma_{i=0}^n (1+t)^{n-i} t^i \binom{n}{i}}{(1+t)^2} = \Sigma_{i=0}^n (1+t)^{n-i-2} t^i \binom{n}{i}$$

and $s = \Sigma_{i=0}^{n-2} \binom{n}{i}$, meaning there at least $\Sigma_{i=0}^{n-2} \binom{n}{i}$ "number" of solutions of $P(x) = 0$ at infinity. Thus there are at most

$$2^n - s = (1+1)^n - \Sigma_{i=0}^{n-2} \binom{n}{i} = n + 1$$

solutions of $P(x) = 0$ in C^n. The system $Q(X) = 0$ is chosen to have the same nonsingular variety at infinity, and this variety stays at infinity as the homotopy progresses from $t = 0$ to $t = 1$. As a result, the infinity solutions stay infinite, the finite solution paths stay finite, and no extraneous paths exist.

This turns out to be a quite typical situation. Even though the system $P(x) = 0$ to be solved has isolated solutions, when viewed in projective space there may be large number of roots at infinity and quite often high-dimensional manifolds of roots at infinity. Extraneous paths are those that are drawn to the manifolds lying at infinity. If $Q(x) = 0$ can be chosen correctly, extraneous paths can be eliminated.

As another example, suppose $P(x) = 0$ is a system of six equations in six unknowns, of which two equations are of third degree and four equations are of second degree. Assume that each of the highest degree terms of each equation involves an x_1 and x_2. Then the system is deficient. The total degree is $3^2 2^4 = 144$, but one can construct a random product homotopy with 65 solution curves that proceed to all solutions of $P(x) = 0$. Implicit in this is the fact that $P(x) = 0$ has at most 65 solutions. Moreover, the generic system with this structure has exactly 65 solutions.

To be more precise, we state the main random product homotopy result. Theorem 2.2 of [10]. Let $V_\infty(Q)$ denote the variety of roots at infinity of $Q(x) = 0$.

Theorem 2.1 *If $V_\infty(Q)$ is nonsingular and contained in $V_\infty(P)$, then Properties 1 and 2 hold.*

Of course, Properties 1 and 2 are not enough. Without starting points, the path-following method cannot get started. Thus $Q(x) = 0$ should also be chosen to be of random product forms, as in (5), which are trivial to solve because of their form.

This result was superseded by the result in [13]. The complex number e_{ij} are chosen random in [10] to ensure Properties 1 and 2. In [13], it was proved that e_{ij} can be any fixed numbers, as long as the complex number c is chosen at random, Properties 1 and 2 still hold. In fact, the result in [13] implies that the start system $Q(x) = 0$ in Theorem 2.1 need not be in product form. It can be any chosen polynomial system whose variety of roots at infinity $V_\infty(Q)$ is nonsingular and contained in $V_\infty(P)$.

Theorem 2.1 in [15] goes one step further. Even if the set $V_\infty(Q)$ of roots at infinity of $Q(x) = 0$ has singularities, if the set is contained in $V_\infty(P)$ *counting multiplicities*, that is, containment in the sense of scheme theory of algebraic geometry, then Properties 1 and 2 still hold. To be more precise, let $I = < \overline{q_1}, \cdots, \overline{q_n} >$ and $J = < \overline{p_1}, \cdots, \overline{p_n} >$ be the homogeneous ideals spanned by homogenizations of q_i's and p_i's respectively. For a point p at infinity, if the local rings I_p and J_p satisfy

$$I_p \subset J_p$$

then Properties 1 and 2 hold. However, this hypothesis can be much more difficult to verify than whether the set is nonsingular. This limits the usefulness of this approach for practical examples.

2.2. m-homogeneous structure

In [19,20], another interesting approach to problem A is developed, using the concept of m-homogeneous structure.

The complex n-space $\mathbf{C}^n$ can be naturally embedded in $\mathbf{P}^n$. Similarly, the space $\mathbf{N} = C^{k_1} \times \cdots \times C^{k_m}$ can be natural embedded in $\mathbf{P} = P^{k_1} \times \cdots \times P^{k_m}$. A point $(y_1, \cdots, y_m)$ in $\mathbf{N}$ with $y_i = (y_1^{(i)}, \cdots, y_{k_i}^{(i)}), i = 1, \cdots, m$ corresponds to a point $(z_1, \cdots, z_m)$ in $\mathbf{P}$ with $z_i = (z_0^{(i)}, \cdots, z_{k_i}^{(i)})$ and $z_0^{(i)} = 1, i = 1, \cdots, m$. The set of such points in $\mathbf{P}$ is usually called the *affine space* in this setting. The points in P with at least one $z_0^{(i)} = 0$ are called the *points at infinity*.

Let f be a polynomial in the n variables $x_1, \cdots, x_n$. If we partition the variables into m groups $y_1 = (x_1^{(1)}, \cdots, x_{k_1}^{(1)})$, $y_2 = (x_1^{(2)}, \cdots, x_{k_2}^{(2)})$, $\cdots$, $y_m = (x_1^{(m)}, \cdots, x_{k_m}^{(m)})$ with $k_1 + \cdots + k_m = n$ and let d_i be the degree of f with respect to y_i(more precisely, to the variables in y_i), then we can define its m-homogenization as

$$\overline{f}(z_1, \cdots, z_m) = (z_0^{(1)})^{d_1} \times \cdots \times (z_0^{(m)})^{d_m} f(y_1/z_0^{(1)}, \cdots, y_m/z_0^{(m)}).$$

$\overline{f}$ is homogeneous with respect to each $z_i = (z_0^{(i)}, \cdots, z_{k_i}^{(i)}), i = 1, \cdots, m$. Here $z_j^{(i)} = x_j^{(i)}$ for $j \neq 0$. Such a polynomial is said to be *m-homogeneous*, and $(d_1, \cdots, d_m)$ is the m-homogeneous degree of f. To illustrate this definition, let us consider the polynomial $p_i(x)$ in (4),

$$\begin{aligned} p_i(x) &= x_1(a_{i1}x_1 + \cdots + a_{in}x_n) + (b_{i1}x_1 + \cdots + b_{in}x_n) + c_i = 0 \\ &= a_{i1}x_1^2 + x_1(a_{i2}x_2 + \cdots + a_{in}x_n) + (b_{i2}x_2 + \cdots + b_{in}x_n) + c_i. \end{aligned}$$

We may let $y_1 = (x_1)$, $y_2 = (x_2, \cdots, x_n)$ and $z_1 = (x_0^{(1)}, x_1)$, $z_2 = (x_0^{(2)}, x_2, \cdots, x_n)$. The degree of $p_i(x)$ is two respect to y_1 and is one with respect to y_2. Hence, its 2-homogeneous is

$$\overline{p_i}(z_1,z_2) = a_{i1}x_1^2x_0^{(2)} + x_1x_0^{(1)}(a_{i2}x_2+\cdots+a_{in}x_n+b_{i1}x_0^{(2)})+$$
$$(x_0^{(1)})^2(b_{i2}x_2+\cdots+b_{in}x_n+c_ix_0^{(1)})$$

which is homogeneous with respect to both $z_1 = (x_0^{(1)}, x_1)$ and $z_2 = (x_0^{(2)}, x_2, \cdots, x_n)$. When the system (4) is viewed in $\mathbf{P}^n$ with the homogenization

$$\begin{aligned}\overline{p_1}(x_0,x_1,\cdots,x_n) =& x_1(a_{11}x_1+\cdots+a_{1n}x_n)+\\ &(b_{11}x_1+\cdots+b_{1n})x_nx_0+c_1x_0^2=0\\ &\vdots\\ \overline{p_n}(x_0,x_1,\cdots,x_n) =& x_1(a_{n1}x_1+\cdots+a_{nn}x_n)+\\ &(b_{n1}x_1+\cdots+b_{nn})x_nx_0+c_nx_0^2=0\end{aligned}$$

its total degree, or the Bézout number , is $d = d_1\cdots d_n = 2^n$. However, when (4) is viewed in $P^1 \times P^n$ with the 2-homogenization

$$\begin{aligned}\overline{p_1}(z_1,z_2) =& a_{11}x_1^2x_0^{(2)} + x_1x_0^{(1)}(a_{12}x_2+\cdots+a_{1n}x_n+b_{11}x_0^{(2)})+\\ &(x_0^{(1)})^2(b_{12}x_2+\cdots+b_{1n}x_n+c_1x_0^{(1)})\\ &\vdots\\ \overline{p_n}(z_1,z_2) =& a_{n1}x_1^2x_0^{(2)} + x_1x_0^{(1)}(a_{n2}x_2+\cdots+a_{nn}x_n+b_{n1}x_0^{(2)})+\\ &(x_0^{(1)})^2(b_{n2}x_2+\cdots+b_{nn}x_n+c_nx_0^{(1)})\end{aligned} \tag{6}$$

the Bézout number d is different. It equals the coefficient of $\alpha_1^1\alpha_2^{n-1}$ in the product $(2\alpha_1+\alpha_2)^n$. That is, $d = 2n$. In general, for the m-homogeneous system

$$\begin{aligned}\overline{p_1}(z_1,\cdots,z_m) &= 0\\ &\vdots\\ \overline{p_n}(z_1,\cdots,z_m) &= 0\end{aligned} \tag{7}$$

in $P^{k_1}\times\cdots\times P^{k_m}$ with $\overline{p_i}$ having m-homogeneous degree $(d_{1,i}, d_{2,i}, \cdots, d_{m,i}), i = 1,\cdots,n$, the Bézout number d of this system is the coefficient of $\alpha_1^{k_1}\times\cdots\times\alpha_m^{k_m}$ in the product

$$(d_{1,1}\alpha_1+\cdots+d_{m,1}\alpha_m)(d_{1,2}\alpha_1+\cdots+d_{m,2}\alpha_m)\cdots(d_{1,n}\alpha_1+\cdots+d_{m,n}\alpha_m).$$

The classical Bézout Theorem says that the system (7) has no more than d isolated solutions, counting multiplicities in $P \equiv P^{k_1}\times\cdots\times P^{k_m}$. Applying to our example in (6), the upper bound of the number of the isolated solutions of (6), in affine space, and at infinity, is $2n$. When solving the original system in (4), we may choose the start system $Q(x) = 0$ in the homotopy

$$H(x,t) = (1-t)cQ(x) + tP(x)$$

to respect the 2-homogeneous structure of $P(x)$. For instance, we may choose $Q(x) = 0$ as

$$\begin{aligned}
q_1(x) &= (x_1 + e_{11})(x_1 + e_{12})(x_2 + \cdots + x_n + e_{13}) \\
q_2(x) &= (x_1 + e_{21})(x_2 + e_{22})(x_2 + e_{23}) \\
&\vdots \\
q_n(x) &= (x_1 + e_{n1})(x_n + e_{n2})(x_n + e_{n3}),
\end{aligned}$$

which has the same 2-homogeneous structure as $P(x)$ with $y_1 = (x_1)$ and $y_2 = (x_2, \cdots, x_n)$. Namely, each $q_i(x)$ has degree two with respect to y_1 and degree one with respect to y_2. It is easy to see that for randomly chosen complex numbers e_{ij}, $Q(x) = 0$ has $2n$ solutions in $\mathbf{C}^n (= \mathbf{C}^1 \times \mathbf{C}^{n-1})$ (thus, no solutions at infinity when viewed in $\mathbf{P}^1 \times \mathbf{P}^{n-1}$). Hence, there are $2n$ paths starting from $2n$ starting points for this choice of homotopy. It is shown in [19] that Properties 1 and 2 hold for all complex number c, except those lying on a finite number of rays starting at the origin. Thus, all solutions of $P(x) = 0$ are found at the end of $n+1$ paths. The number of extraneous paths $2n - (n+1) = n - 1$ is far less than the number of extraneous paths $2^n - n - 1$ by using the classical homotopy with $Q(x) = 0$ in (3). More precisely, we state the main Theorem in [19].

Theorem 2.2 *Let $Q(x)$ be a system of equations chosen to have the same m-homogeneous form as $P(x)$. Assuming that $Q(x) = 0$ has exactly the Bézout number, d, of nonsingular solution, and define*

$$H(x,t) = (1-t)cQ(x) + tP(x)$$

where $t \in [0,1]$ and $c \in C$. Then for all but a finite number of angles, θ, if $c = re^{i\theta}$ for some positive $r \in R$, Properties 1 and 2 hold.

Again, there are still $n-1$ extraneous paths in the example we did above because, even viewed $\mathbf{P}^1 \times \mathbf{P}^{n-1}, P(x)$ has zeros at infinity. To go one step further, in addition to starting the same 2-homogeneous structure of $P(x)$, we may choose the start system $Q(x) = 0$ to have the same nonsingular variety of roots of $P(x) = 0$ at infinity. The choice of $Q(x) = 0$ in (5) not only shares the same 2-homogeneous structure of $P(x)$ with $y_1 = (x_1)$ and $y_2 = (x_2, \cdots, x_n)$ but also, when viewed in $\mathbf{P}^1 \times \mathbf{P}^{n-1}$ with $z_1 = (x_0^{(1)}, x_1)$, and $z_2 = (x_0^{(2)}, x_2, \cdots, x_n)$, has the same nonsingular variety of roots of $P(x) = 0$ at infinity as $\overline{P}(z_1, z_2)$ defined by $x_1 = x_0^{(2)} = 0$. The system $Q(x) = 0$ also has $n+1$ solutions in C^n, and there are no extraneous paths. In general, it can be shown [15, 20] that if $Q(x) = 0$ in

$$H(x,t) = (1-t)cQ(x) + tP(x)$$

is chosen to have the same m-homogeneous form as $P(x)$ and assume that the set of zeros $V_\infty(Q)$ of $Q(x)$ at infinity is nonsingular and contained in $V_\infty(P)$, then all but a finite number of angle, θ, if $c = re^{i\theta}$ for some positive $r \in R$, Properties 1 and 2 hold.

The zeros of an m-homogeneous polynomial system $\overline{P}(z_1,\cdots,z_m)$ at infinity in $\mathbf{P}^{k_1}\times\cdots\times\mathbf{P}^{k_m}$ may sometimes be difficult to obtain. Nevertheless, the choice of $Q(x)=0$ in Theorem 2.2, assuming no zeros at infinity regardless of the structure of the zeros at infinity of $P(x)$, can still reduce the number of extraneous paths drastically in many solutions for simply sharing the same m-homogeneous structure of $P(x)$.

The usefulness of the methods developed so far for problem A is restricted to application on a case-by-case basis. The challenge is, in a specific case, to find a $Q(x)$ that is simple to solve (Property 0) and also produces few extraneous paths.

3. Methods for Problem B

The situation for Problem B is different. A method called the cheater's homotopy [11,12](a similar procedure can be found in [21]) has been developed which is, in some sense, an optimum solution procedure. Problem B asks that the system $P(c,x)=0$ be solved for several different values of the coefficient c. In other words, we think of $P(c,x)=0$ as a system with some structure or sparsity.

The idea of the method is to theoretically establish Properties 1 and 2 by deforming from a sufficiently generic system (in a precise sense to be given later) and then to "cheat" on Property 0 by using a preprocessing step. The amount of computation of preprocessing step is large, but is amortized among the several solving characteristic of Problem B.

We begin with an example. Let $P(x)$ be the system

$$\begin{aligned} p_1(x) &= x_1^3x_2^2 + c_1x_1^3x_2 + x_2^2 + c_2x_1 + c_3 = 0 \\ p_2(x) &= c_4x_1^4x_2^2 - x_1^2x_2 + x_2 + c_5 = 0 \end{aligned} \tag{8}$$

This is a system of two polynomial equations in two unknowns x_1, x_2. We want to solve Problem B, meaning we want to solve the system of equations several times, for various specific choice of $c=(c_1,\cdots,c_5)$.

It turns out that for any choice coefficients c, system (8) has no more than 10 isolated solutions. More precisely, there is an open dense subset S of C^5 such that for c belonging to S, there are 10 solutions of (8). Moreover, 10 is the upper bound for the number of isolated solutions for all c in C^5. The total degree of the system is $6\cdot 5=30$, meaning that if we had taken a generic system of two polynomials in two variables of degree 5 and 6. there are would be 30 solutions. Instead, all but five of the possible coefficients are fixed in (8), resulting in fewer than 30 solutions. Thus (8), with any choice of c, is a deficient system.

The classical homotopy continuation method produces $d=30$ paths, beginning at 30 trivial starting points. Thus, there are (at least) 20 extraneous paths.

The cheater's homotopy continuation approach to solve Problem B is to begin by solving (8) with *randomly-chosen* complex coefficient $c^* = (c_1^*, \cdots, c_5^*)$. Call X^* the set of 10 solutions. No work is saved there, since 30 paths need to be followed, and 20 paths are wasted. However, the 10 elements of the set X^* are the seeds for the remainder of the process. In the future, for each choice of the coefficients $c = (c_1, \cdots, c_5)$ for which the system (8) needs to be solved, we use the homotopy continuation method to follow a straight-line homotopy from the system with the coefficients c_* to the system with coefficients c. We follow the 10 paths beginning at the 10 elements of X^*. Thus Property 0, that of having trivial-available starting points, is satisfied. The fact that Properties 1 and 2 are also satisfied is the content of Theorem 3.1 below. Thus, for each fixed c, all (or fewer) 10 isolated solutions of (8) lie at the end of 10 smooth homotopy paths beginning at the seeds in X^*. After the foundational step of finding the seeds, the complexity of all further solving of (8) is proportional to the number of solutions 10, rather than the total degree 30.

Furthermore, this method, unlike the method for problem A, requires no a priori analysis of the system. The first preprocessing step of finding the seeds establishes a sharp theoretical upper bound on the number of isolated solutions as a by-product of the computation; further solving of the system use the optimal number of paths to be followed.

We earlier characterized a successful homotopy continuation method as having three properties: triviality, smoothness, and accessibility. Given an arbitrary system of polynomial equations, such as (8), it is not too hard (through generic perturbations) to find a family of systems which has the last two properties. The problem is that one member of this family must be trivial to solve, or the path-following cannot get started. The idea of the cheater's homotopy is simply to "cheat" on this part of the problem, and run a preprocessing step(the computation of the seeds of X^*) which gives us the triviality property in a roundabout way. Thus the name, the "cheater's homotopy".

A statement of the theoretical result we need follows. Let

$$\begin{aligned} p_1(c_1, \cdots, c_M, x_1, \cdots, x_n) &= 0 \\ &\vdots \\ p_n(c_1, \cdots, c_M, x_1, \cdots, x_n) &= 0 \end{aligned} \tag{9}$$

be a system of polynomial equations in the variables $c_1, \cdots, c_M, x_1, \cdots, x_n$. For each choice of $c = (c_1, \cdots, c_M)$ in C^M, this is a system of polynomial equations in the variables $x_1, \cdots, x_n$. Let d be the total degree of the system for a generic choice of c

Theorem 3.1 *Let c belong to C^M. There exists an open dense full-measure subset U of C^{n+M} such that for $(b_1^*, \cdots, b_n^*, c_1^*, \cdots, c_M^*) \in U$, the following holds:*

(a) The set X^* of solutions $x = (x_1, \cdots, x_n)$ of

$$\begin{aligned} q_1(x_1,\cdots,x_n) &= p_1(c_1^*,\cdots,c_M^*,x_1,\cdots,x_n) + b_1^* = 0 \\ &\vdots \\ q_n(x_1,\cdots,x_n) &= p_n(c_1^*,\cdots,c_M^*,x_1,\cdots,x_n) + b_n^* = 0 \end{aligned} \tag{10}$$

consists of d_0 isolated points, for some $d_0 \leq d$.
(b) The smoothness and accessibility properties hold for the homotopy

$$H(x,t) = P((1-t)c_1^* + tc_1, \cdots, (1-t)c_M^* + tc_M, x_1, \cdots, x_n) + (1-t)b^* \tag{11}$$

where $b^* = (b_1^*, \cdots, b_n^*)$. It follows that every solution of $P(x) = 0$ is reached by a path beginning at a point of X^*.

A proof of Theorem 3.1 can be found in [11]. The theorem is used as part of the following procedure. Let $P(c,x) = 0$ as in (9) denote the system to solved for various values of the coefficients c.

Cheater's homotopy Procedure.

Step 1. Choose complex number $(b_1^*, \cdots, b_n^*, c_1^*, \cdots, c_m^*)$ at random, and use the classical homotopy continuation method to solve $Q(x) = 0$ in (10). Let d_0 denote the number of solutions found.(This number is bounded above by the total degree d.) Let X^* denote the set of d_0 solutions.

Step 2. For each new choice of coefficients $c = (c_1, \cdots, c_M)$, follow the d_0 paths defined by $H(x,t) = 0$ in (11), beginning at the points of in X^*, to find all solutions of $P(c,x) = 0$.

In Step 1 above, when complex numbers $(c_1^*, \cdots, c_M^*)$ are chosen at random, using the classical homotopy continuation methods to solve $Q(x) = 0$ in (10) may sometimes be a heavy work by itself. It is desirable that those numbers do not have to be chosen at random. For the sake of illustration, consider the linear system

$$\begin{aligned} c_{11}x_1 + \cdots + c_{1n}x_n &= b_1 \\ &\vdots \\ c_{n1}x_1 + \cdots + c_{nn}x_n &= b_n \end{aligned} \tag{12}$$

which may be considered as a polynomial system with degree one of each equation. For randomly chosen $c'_{ij}s$, (12) has a unique solution which is not available right away. However, if we choose $c_{ij} = \delta_{ij}$, the solution is obvious.

For this purpose, an alternative is suggested in [16]. When a system $P(c,x) = 0$ with a particular parameter c^0 is solved, then for any parameter $c \in C^M$ consider the nonlinear homotopy

$$H(a,x,t) = P((1 - [t - t(1-t)a])c^0 + (t - t(1-t)a)c, x) = 0. \tag{13}$$

It is proved in [16] that for randomly chosen complex number a the solution paths of $H(a,x,t) = 0$ in (13), emanating from the solutions of $P(c^0, x) = 0$ will reach the isolated solutions of $P(c,x) = 0$ under the natural assumption that for generic c, $P(c,x)$ has the same number of solutions in C^n.

The most important advantage of the homotopy in (13) is that the parameter c^0 of the start system $P(c^0,x) = 0$ need not be chosen at random as long as it is chosen for which $P(c^0,x) = 0$ has the same number of solutions as $P(c,x) = 0$ for generic c. Therefore, in some situations, when the solutions of $P(c,x) = 0$ is easily available for certain parameter c_0, the system $P(c^0,x) = 0$ may be used as the start system in (13) and the effort of solving $P(c,x) = 0$ for a randomly chosen c as a start system is saved.

To finish, we give a more non-trivial example of the use of the procedure described in this section.

Consider the indirect position problem for revolute-joint kinematic manipulators. Each joint represents a one-dimensional choice of parameters, namely the angular position of the joint. If all angular positions are known, then of course the position and orientation of the end of the manipulator (the hand) are determined. The indirect position problem is the inverse problem: given the desired position and orientation of the hand, find a set of angular parameters for the (controllable) joints which will put the hand in the desired state.

The indirect position problem for six joints is reduced to a system of eight nonlinear equations in eight unknowns in [22]. The coefficients of the equations depend on the desired position and orientation, and solution of the system (an eight-vector) represents the sines and cosines of the angular parameters. Whenever the manipulator's position is changed, the system need to be resolved with new coefficients. The equations are too long to repeat here (see the appendix of [22]); suffice to say that it is a system of eight degree-two polynomial equations in eight unknowns which is quite deficient. The total degree of the system is $2^8 = 256$, but there are at most 32 isolated solutions.

The nonlinear homotopy (13) requires only 32 paths to solves the system with different parameters([14, 16]). The system contains 26 coefficients, and a specific set of coefficients is chosen for which the system has 32 solutions. For subsequent solving of the system, for any choice of the coefficients $c_1, \cdots, c_{26}$, all solutions are found at the end of exactly 32 paths, by using nonlinear homotopy in (13) with randomly chosen complex number a.

4 Numerical Considerations

4.1 Projective Coordinates

As described in Section 1, the solution paths of (2) which do not proceed to a solution of $P(x)=0$ in C^n diverge to infinity, a very poor state of affairs for numerical method. However, there is a simple idea from classical mathematics which improves the situation. If the system (2) is viewed in P^n, the diverging paths are simply proceeding to a "point at infinity" in projective space. Since projective space is compact, we can force all paths, including the extraneous ones, to have finite length by using projective coordinates.

Instead of $P(x)=0$ consider the system of $n+1$ equations in $n+1$ unknowns

$$\begin{aligned} \overline{p_1}(x_0,\cdots,x_n) &= 0 \\ &\vdots \\ \overline{p_n}(x_0,\cdots,x_n) &= 0 \\ a_0x_0+\cdots+a_nx_n-1 &= 0 \end{aligned}$$

where $a_0,\cdots,a_n$ are complex numbers. Using the classical homotopy continuation procedure to follow the $d=d_1\cdots d_n$ paths of this system of $n+1$ equations in $n+1$ unknowns. For almost every choice $a_0,\cdots,a_n$, the paths stay in $\mathbf{C}^{n+1}$. It remains only to throw out the solutions which were at infinity for the original problem (those with $x_0=0$). Of the remaining solutions with $x_0\neq 0$, it is easy to see that $x=(x_1/x_0,\cdots,x_n/x_0)$ is the corresponding solution of $P(x)=0$.

A similar technique is described in [20], where it is called a "projective transformation". It differs from the above in the following way. Instead of increasing the size of the problem from n by n to $n+1$ by $n+1$ as above, they implicitly consider solving the last equation for x_0 and substituting in the other equations, essentially retaining n equations in n unknowns. Then the chain rule is used for purpose of the Jacobian calculations which are necessary for the path following. In many cases, it seems that this may create extra work. Suppose, for example, that the tenth equation in the system is $p_{10}(x)=x_7^3-x_1x_2$. The homogeneous version is $\overline{p_{10}}(x)=x_7^3-x_0x_1x_2$. Since now x_0 is considered a function of all other variables, the partial derivative of p_{10} with respect to every variables is suddenly nonzero. This result is added computation for each Jacobian evaluation, and is particularly problematic if the original problem is large and/or sparse.

4.2 Real Homotopy

Most practical polynomial systems in application consist of polynomials with real coefficients, and most often the only desired solutions are real solutions. This suggests the usage of the real homotopies. That is, when the coefficients of the target polynomial system $P(x)=0$ we want to solve are all real, we may choose the start system $Q(x)=0$ with real coefficients, making the homotopy $H(x,t)=0$ consisting of real polynomial system for each t. Thus, for fixed t, if x is a solution of $H(x,t)=0$ so is its conjugate $\overline{x}$. Accordingly, a major advantage of the real homotopy is that when a complex homotopy path $(\overline{x}(s), t(s))$ can be obtained as a by-product without any further computations. On the other hand, although the homotopy $H(x,t)$ is still a map from $C^n \times [0,1]$ to C^n, when a real homotopy path is traced, we may consider $H(x,t)$ as a map from $R^n \times [0,1]$ to R^n, and hence the computation of the following the real path may stay in real space, and use real arithmetics. In this way, a considerable amount of computation is reduced.

There are numerous computational problems associated with the path following algorithms of real homotopies. In particular, when real homotopies are used, different from the complex homotopy, bifurcation of some of the homotopy paths is inevitable. Hence, efficient algorithms must be developed to identify the bifurcation points and to follow the path after bifurcation. It is easy to see that bifurcations can only occur at *turning points*, points (x^*, t^*) for which $\dot{t} = 0$ and $H_x(x^*, t^*)$ is singular. To identify the bifurcation point, let $p_0 = (x_0, t_0)$ be a point on the solution path Γ with $\dot{t}(p_0) > 0$. After standard Euler prediction with step size h_0 and Newton corrections, we obtain a point $p_1 = (x_1, t_1)$ on Γ. When tangent vector $(\dot{x}, \dot{t})$ is calculated at p_1 and $\dot{t}(p_1) < 0$, apparently a turning point $p^* = (x^*, t^*)$ exists in this situation.

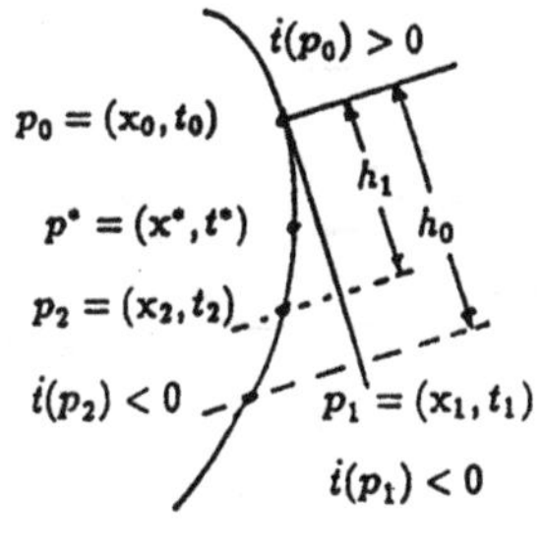

Figure 2:

To identify $\mathbf{p}^*$, we take the following procedure:

1. Let h_1 be the solution of the equation

$$\frac{h}{h_0}\dot{t}(\mathbf{p}_0) + \frac{h_0 - h}{h_0}\dot{t}(\mathbf{p}_1) = 0.$$

Taking Euler prediction at $\mathbf{p}_0$ with step size h_1 followed by Newton corrections, we obtain a new point $\mathbf{p}_2$ on Γ.

2. If $\dot{t}(\mathbf{p}_2) > 0$, we replace $\mathbf{p}_0$ by $\mathbf{p}_2$ and replace h_0 by the real part of the inner product of $\mathbf{p}_1 - \mathbf{p}_2$ and the unit tangent vector at $\mathbf{p}_2$. If $\dot{t}(\mathbf{p}_2) < 0$, we replace $\mathbf{p}_1$ by $\mathbf{p}_2$ and h_0 by h_1.

3. Repeat Step 1, until $\dot{t}(\mathbf{p}_2)$ is small enough. Then, $\mathbf{p}_2$ will be taken as a bifurcation point $\mathbf{p}^* = (\mathbf{x}^*, t^*)$.

When the bifurcation point p^* is identified, in order to follow the bifurcation branches, tangent vectors of the branches need to be characterized. It turns out that for the following special kind of turning points the bifurcation phenomenon is rather simple.

Definition 4.1 *A singular point* $(x^*,t^*) \in C^n \times [0,1]$ *is said to be a* quadratic *turning point of* $H(x,t)=0$ *if*
1 $Rank_R H_x(x^*,t^*) = 2n-2$
2 $Rank_R[H_x(x^*,t^*), H_t(x^*,t^*)] = 2n-1$
3 For $0 \neq y \in C^n$ *with* $H_x(x^*,t^*)y = 0, Rank_R(H_x(x^*,t^*), H_{xx}(x^*,t^*)yy) = 2n$.
Here $Rank_R$ *denote the real rank.*

Proposition 4.1 *Let* (x^*,t^*) *be a quadratic turning point. Then, there are only two branches of solution paths* Γ *and* Γ'*passing through* (x^*,t^*)*. If* ϕ *is the tangent vector of the path* Γ *at* (x^*,t^*)*, then the tangent vector of* Γ' *at* (x^*,t^*) *is in the direction of* $i\phi$*. (See Figure 3.)*

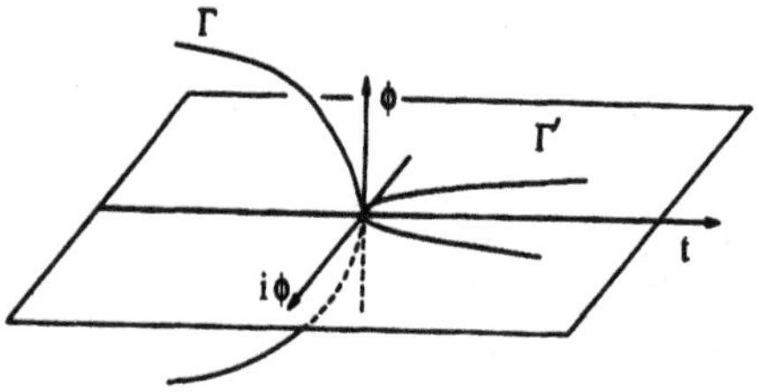

Figure 3:

(When Γ *is a real path, the assertion of the proposition can be considered a special case in [2, 7]. The most general version, where* Γ *and* x^* *may both be complex, was proved in [17].)*

To follow the bifurcation branch Γ' at a quadratic turning point, we consider the following three situations:

1. Γ *is a real path.*

Then ϕ is real and $i\phi$ is pure imaginary. Apparently, the bifurcation branch Γ' consists of a complex path and its complex conjugacy. We need only to follow one of them with tangent vector $i\phi$ or $-i\phi$.

2. Γ *is a complex path and* (x^*,t^*) *is real.*

Then, Γ consists of complex conjugate pairs (x,t) and $(\bar{x},t)$ for each $t < t^*$. The tangent vector at (x^*, t^*)

$$\phi = lim_{s_1 - s_2 \to 0} \frac{(x(s_1), t(s_1)) - (x(s_2), t(s_2))}{s_1 - s_2}, \ where \ x(s_2) = \overline{x}(s_1), t(s_1) = t(s_2)$$

is clearly pure imaginary. Hence, $i\phi$ is real. Consequently, the bifurcation branch Γ' consists of two real paths. We may follow them in real space $R^n \times [0,1]$ with real tangent vectors $i\phi$ and $-i\phi$ respectively.

3. Γ *is a complex path and* x^* *is complex.*

The bifurcation branch Γ', in this case, consists of two complex solution paths. They are not conjugate to each other. We may follow them with tangent vectors $i\phi$ and $-i\phi$ respectively.

It was conjectured in [3] that, *generically*, real homotopies contain no singular points other than a finite number of quadratic turning points. A paper by Verlinden and Haegemans [23] asserts that this conjecture is true but the supposed proof has a gap, and the authors agreed this gap exists. A review in *Mathematical Review*(MR.91e, p. 2783,#65071) reported this error. Recently, this conjecture is proved in [17] from a completely different approach.

Modern scientific computing is marked by the advent of vector and parallel computers and the search for algorithms that are to a large extent parallel in nature. A great advantage of the homotopy continuation algorithm for solving polynomial systems is that it is to a large degree parallel, in the sense that each homotopy path is followed independent of the others. This nature parallelism makes the method an excellent candidate for a variety of architectures. In this respect, it stands in contrast to the highly serial Groebner bases method and the μ-resultant method. We expect a very high level of performance of our algorithm on multiprocessors with shared or distributed memory.

References

[1] E. Allgower and K. Georg, "Numerical Continuation Methods", *Springer-Verlag, New York, 1990.*

[2] E. L. Allgower, "Bifurcations arising in the calculation of critical points via homotopy methods", in *Numerical Methods for Bifurcation problems,* Proceedings of the conference at the University of Dortmund, T. Kupper, H. D. Mittelman, and H. Weber eds., Birkhauser Verlag, Basel, 1984. 15-28.

[3] P. Brunovsky and Meravy, "Solving systems of polynomial systems by bounded and real homotopy", *Numer. Math.*, Vol.43, (1984), 397-418.

[4] S. N. Chow, J. Mallet-Paret and J. A. Yorke, "A homotopy method for locating

all zeros of a system of polynomials", *Functional Differential Equations and Approximation of Fixed Points, Peitgen, H.O. and Walther, H.O. ed. Springer, Leture Notes in Mathematics* No. 730 (1979), 77-88.

[5] F. J. Drexler, "Eine Methode Zur Berechnung Samlischer Losüngen von Polynomgleichungssystemen", *Numer. Math.*, 29 ,(1977), 45-58.

[6] C. B. Garcia and W. I. Zangwill, "Finding all solutions to polynomial systems and other systems of equations", *Math. Programming* 16 (1979), 159-176.

[7] M. E. Henderson and H. B. Keller, "Complex bifurcation from real paths", *SIAM J. Appl Math.,* Vol.50, No.2, (1990), 460-482.

[8] T. Y. Li, " On Chow Mallet-Paret and York homotopy for solving systems of polynomials", *Bulletin of Institute of Mathematics, Academica Sinica 11* (1983), 433-437.

[9] T. Y. Li, T. Sauer and J. Yorke, "Numerical solution of a class of deficient polynomial systems", *SIAM J. Numer. Anal.* Vol.24, No.2(1987), 435-451.

[10] T. Y. Li, T. Sauer and J. A. Yorke, "The random product homotopy and deficient polynomial systems", *Numer. Math.*, 51, (1987), 481-500.

[11] T. Y. Li, T. Sauer and J. A. Yorke, "The cheater's homotopy: An efficient procedure for solving systems of polynomial equations", *SIAM J. Numer. Anal.,* Vol.26, No.5 (1989), 1241-1251.

[12] T. Y. Li, T. Sauer and J. A. Yorke, "Numerically determining solutions of systems of polynomial equations", *Bull. A. M. S.* 18 (1988), 173-177.

[13] T. Y. Li, T. Sauer, " A simple homotopy for solving deficient polynomial systems," *Japan J. Appl. Math.*, 6 (1989), 409-419. An efficient procedure for solving systems of polynomial equations," *SIAM J. Numer. Anal.*,Vol.26, No.5 (1989), 1241-1251.

[14] T. Y. Li and X. Wang, "A homotopy for solving the kinematics of the most general six-and-five-degree of freedom manipulators", *Proc. of ASME Conference on Mechanisms,* D1-Vol. 25, (1990) 249-252.

[15] T. Y. Li and X. Wang, "Solving deficient polynomial system with homotopies which keep the subschemes at infinity invariant", *Math Comp.* Vol.56, No.194, (1991), pp.693-710.

[16] T. Y. Li and X. Wang, "Nonlinear homotopies for solving deficient polynomial system with parameters", to appear: *SIAM J. of Numer.*

[17] T. Y. Li and X. Wang, "Solving real polynomial systems with real homotopies", a preprint.

[18] A. P. Morgan, "A homotopy for solving polynomial systems" *Applied Math. Comp.*

18 (1986), 87-92.

[19] A. P. Morgan and A. J. Sommese, "A homotopy for solving general polynomial systems that respect m-homogeneous structures" *Appl. Math. Comp.* 24 (1987), 101-113.

[20] A. P. Morgan and A. J. Sommese, " Computing all solutions to polynomial systems using homotopy continuation" *Appl. Math. Comp.* 24 (1987), 115-138.

[21] A. P. Morgan and A. J. Sommese, " Coefficient parameter polynomial continuation" *Appl. Math. Comp.* 29 (1989), 123-160.

[22] L.-W. Tsai and A. Morgan, "Solving the kinematics of the most general six-and-five-degree-of-freedom manipulators by continuation methods", ASME J. Mechanisms, Transmissions and Automation in Design, 107 (1985): 48-57.

[23] P. Verlinden and Haegemans, "Real homotopy for solving systems of real polynomial equations", *Bull. Soc. Math. Belg.* 41 (1989), 325-338.

[24] A. H. Wright, "Finding all solutions to a system of polynomial equations," *Math. Comp.* 44 (1985), 125-133.

[25] W. Zulener, " A simple homotopy method for determining all isolated solutions to polynomial systems," *Math. Comp.* 50 (1985), 167-177.

Computer Mathematics
Proc. of the Special Program at
Nankai Institute of Mathematics
January 1991 – June 1991

The Random Product Homotopy for Solving Polynomial Systems in $C^{k_1} \times C^{k_2} \times \cdots \times C^{k_m}$

Yu Bo and Feng Guo-chen

Computer Center, Jilin University
Changchun, 130023, P.R.China

Abstract

In this paper, we construct a homotopy for solving general polynomial systems in $C^{k_1} \times C^{k_2} \times \cdots \times C^{k_m}$ which are highly deficient in C^n $(n = k_1 + k_2 + \cdots + k_m)$. It works with probability one, and is performed in C^n or C^{n+1}. The start system can be solved easily.

1. Introduction

Let C^{k_j} $(1 \leq j \leq m)$ be complex spaces of dimension k_j respectively, and $X = (X_1, X_2, \cdots, X_m) \in C^{k_1} \times C^{k_2} \times \cdots \times C^{k_m}$, where $X_j = (x_{j1}, x_{j2}, \cdots, x_{jk_j}), k_1 + k_2 + \cdots + k_m = n$. A polynomial in n variables over C is called a polynomial in $C^{k_1} \times C^{k_2} \times \cdots \times C^{k_m}$, if it is represented in the following form

$$p(X) = \sum_{|\gamma_1| \leq d_1, \cdots, |\gamma_m| \leq d_m} \alpha_{\gamma_1 \gamma_2 \cdots \gamma_m} X_1^{\gamma_1} X_2^{\gamma_2} \cdots X_m^{\gamma_m} \tag{1.1}$$

where $\gamma_j = \gamma_{j1}\gamma_{j2} \cdots \gamma_{jk_j}$ are multi-indices, $X_j^{\gamma_j} = x_{j1}^{\gamma_{j1}} x_{j2}^{\gamma_{j2}} \cdots x_{jk_j}^{\gamma_{jk_j}}$, $|\gamma_j| = \gamma_{j1} + \gamma_{j2} + \cdots + \gamma_{jk_j}$, and for every $j (1 \leq j \leq m)$ there is at least one $\alpha_{\gamma_1 \gamma_2 \cdots \gamma_m} \neq 0$, such that $|\gamma_j| = d_j$. We call $(d_1, d_2, \cdots, d_m)$ the m-degree of p, and d_j the degree of p with respect to X_j.

In algebraic geometry (cf. Shafarevich [23], Ch. IV, Section 2), the Bezout theorem in $C^{k_1} \times C^{k_2} \times \cdots \times C^{k_m}$ (called m-Bezout theorem in this paper for convenience) tells us the upper bound of the number of the common zeros of a polynomial system in $C^{k_1} \times C^{k_2} \times \cdots \times C^{k_m}$.

m-Bezout Theorem (cf. [20]). Let $p_1, p_2, \cdots, p_n$ be polynomials in $C^{k_1} \times C^{k_2} \times$

$\cdots \times C^{k_m}$, and the m-degree of p_i be $(d_{i1}, d_{i2}, \cdots, d_{im})(i = 1, 2, \cdots, n)$. Then the polynomial system $P = (p_1, p_2, \cdots, p_n)$ has no more than D_m isolated common zeros, where D_m, the m-Bezout number of P, is the coefficient of $t_1^{k_1} t_2^{k_2} \cdots t_m^{k_m}$ in the combinatorial product

$$\prod_{i=1}^{n} \sum_{j=1}^{m} d_{ij} t_j$$

If $m = 1$, the m-Bezout theorem is the Bezout theorem in C^n, and the m-Bezout number is the Bezout number (the product of all degrees of polynomials in the system). In many cases, the m-Bezout number is a better upper bound of the number of isolated zeros than that given by the Bezout number. We take the ordinary algebraic eigenvalue problem as an example.

Example 1.1.

$$\begin{aligned} Ax &= \lambda x \\ c^\tau x &= 1 \end{aligned} \tag{1.2}$$

where A is an $n \times n$ matrix, $x \in C^n$, and c is a complex n-vector. As a polynomial system in $\lambda, x_1, x_2, \cdots, x_n$, the Bezout number of (1.2) is 2^n. If we let $X_1 = (\lambda), X_2 = (x_1, x_2, \cdots, x_n)$, then the 2-Bezout number D_2 of (1.2) is the coefficient of $t_1 t_2^n$ in

$$(t_1 + t_2)^n t_2$$

that is, $D_2 = n$.

For a given system in C^n, we will view it as a system in $C^{k_1} \times C^{k_2} \times \cdots \times C^{k_m}$ for some m and a partition $(X_1, X_2, \cdots, X_m)$ of the variables which makes the m-Bezout number as small as possible.

The homotopy method for solving polynomial systems, presented by Garcia and Zangwill [8] and Drexler [6] independently, is to construct a homotopy map

$$H(x,t) : C^n \times [0,1] \longrightarrow C^n$$

such that $Q(x) = H(x,0)$ (*start system*) is a trivial polynomial system, and $P(x) = H(x,1)$ (*target system*) is the polynomial system to be solved. If $H(x,t)$ is constructed correctly, all isolated zeros of $P(x)$ can be obtained by following smooth paths starting from zeros of $Q(x)$ using standard numerical techniques (cf. [1]).

The typical homotopy maps suggested by Li [11] or Zulehner [27] (cf. also [3], [4], [19], [26]) generate the Bezout number of paths to be followed.

By a deficient system (cf. [12],[13]) we mean the system which has fewer isolated solutions than its Bezout number. In many practical cases, the systems are all deficient. Solving deficient systems using the original homotopy method may cause wasted computation. Some homotopy methods have been presented for solving deficient polynomial systems (see, [12]-[17],[20],[21]). In [12], [13], the random product homotopy was presented.

If the m-Bezout number of a system for some m and a partition of the variables is smaller than its Bezout number, then the system is deficient, and it is usually highly deficient (the system (1.2), for example).

Morgan and Sommese [20] (cf. also [21] and Li and Wang [17] for more information) have pointed out that, if one can construct a homotopy, whose start system has the same m-degree as the target system and has exactly D_m distinct isolated solutions, then one need only to trace D_m paths. But they have not constructed a start system explicitly.

In this paper, by viewing the polynomial system as a special kind of deficient system in C^n, we construct a homotopy (we use the terms *random product homotopy* used by Li, Sauer and Yorke [13]) for general systems in $C^{k_1} \times C^{k_2} \times \cdots \times C^{k_m}$. Its start system can be solved easily. Moreover, instead of C^{n+m}, we can perform the operation in C^n or C^{n+1}.

2. Construction of the homotopy

Let $p_1, p_2, \cdots, p_n$ be polynomials in $C^{k_1} \times C^{k_2} \times \cdots \times C^{k_m}$. We denote $P = (p_1, p_2, \cdots, p_n)$. Let the m-degree of p_i be $(d_{i1}, d_{i2}, \cdots, d_{im_i})$.

To solve

$$P(X) = 0$$

we construct the homotopy map as follows:

$$H(X,t) = (1-t)cQ(X) + tP(X),$$

where $Q = (q_1, q_2, \cdots, q_n)$,

$$q_i(X) = \prod_{j=1}^{m}\prod_{l=1}^{d_{ij}}(\sum_{k=1}^{k_j} a^i_{jk}x_{jk} - b^i_{jl}), \quad i = 1, 2, \cdots, n.$$

Let $a^i_j = (a^i_{j1}, a^i_{j2}, \cdots, a^i_{jk_j})$ be a row vector. For any index sets $I_j = (i_1, i_2, \cdots, i_{k_j})$, $L_j = (l_1, l_2, \cdots, l_{k_j})$, let $A(I_j)$ be a $k_j \times k_j$ matrix whose sth row is $a^{i_s}_j$ and $B(I_j, L_j)$ be a k_j-vector whose sth element is $b^{i_s}_{jl_s}$.

$q_i(X)$ is also a polynomial in $C^{k_1} \times C^{k_2} \times \cdots \times C^{k_m}$ of m-degree $(d_{i1}, d_{i2}, \cdots, d_{im})$.

To solve $Q(X) = 0$, we need only to solve the following sets of linear systems:

$$\begin{aligned} A(I_1)X_1 &= B(I_1, L_1) \\ A(I_2)X_2 &= B(I_2, L_2) \\ &\cdots \\ A(I_m)X_m &= B(I_m, L_m) \end{aligned}$$

for all possible combinations of I_j, L_j such that $I_1 \bigcup I_2 \bigcup \cdots \bigcup I_m = \{1, 2, \cdots, n\}$.

We state the main theorem to conclude this section.

Theorem 2.1. If $A(I_j)$ is nonsingular for any $1 \le j \le m$ and any set of indices $I_j = (i_1, i_2, \cdots, i_{k_j})$ and the solution of the linear system

$$A(I'_j)X_j = B(I'_j, L'_j)$$

is different from the solution of the linear system

$$A(I_j)X_j = B(I_j, L_j)$$

for $(I'_j, L'_j) \ne (I_j, L_j)$, then there exists a finite subset S of $[0,2\pi)$ such that for $\theta \notin S$ and $c = re^{i\theta}$ ($r \ne 0$ real),

$$H^{-1}(0) = \{(X,t) | H(X,t) = 0, t \in [0,1)\}$$

consists of D_m smooth curves parameterized by t. Every isolated solution of $P(X) = 0$ can be reached by such a curve starting from a zero of $Q(X)$.

If $P(X) = 0$ has exactly D_m isolated solutions, then the operation can be performed in C^n, otherwise it can be done in C^{n+1} (see [3],[16],[18],[26]).

Remark 2.2. If $d_{ij} > 1$ for some $1 \le i \le n, 1 \le j \le d_{im}$, then $Q(X)$ will no longer satisfy the conditions in [13].

3. Proof of the Theorem

Let F be a polynomial in the variables $Z_1 = (x_{10}, x_{11}, \cdots, x_{1k_1}), Z_2 = (x_{20}, x_{21}, \cdots, x_{2k_2}), \cdots, Z_m = (x_{m0}, x_{m1}, \cdots, x_{mk_m})$. F is called m-homogeneous if F is homogeneous with respect to Z_j for $j = 1, 2, \cdots, m$.

Lemma 3.1 (Van Der Waerden [24], 80, Ex.4). Let $F_1, F_2, \cdots, F_r$ be m-homogeneous polynomials in $Z_1, Z_2, \cdots, Z_m$, having indeterminate coefficients $c_1, c_2, \cdots, c_N$. Then there exists a set $g_1, g_2, \cdots, g_t$ of homogeneous polynomials in $c_1, c_2, \cdots, c_N$, with integer coefficients, such that for any set of special values of $c_1, c_2, \cdots, c_N \in C, F_1, F_2, \cdots, F_r$ have a nontrivial common zero ($Z_j \ne 0, j = 1, 2, \cdots, m$) if and only if $(c_1, c_2, \cdots, c_N)$ is a common zero of $g_1, g_2, \cdots, g_t$.

Let $P(X_1, X_2, \cdots, X_m)$ be a polynomial in $C^{k_1} \times C^{k_2} \times \cdots \times C^{k_m}$ of m-degree $(d_1, d_2, \cdots, d_m)$. We define its m-homogenization as

$$\tilde{P}^m(Z_1, Z_2, \cdots, Z_m) = x_{10}^{d_1} x_{20}^{d_2} \cdots x_{m0}^{d_m} P(X_1/x_{10}, X_2/x_{20}, \cdots, X_m/x_{m0})$$

$\tilde{P}^m$ is an m-homogeneous polynomial in $Z_j = (x_{j0}, x_{j1}, \cdots, x_{jk_j}), j = 1, 2, \cdots, m$.

For any collection of $1 \le j_1, j_2, \cdots, j_r \le m, \sigma = (j_1, j_2, \cdots, j_r)$, let $\tilde{P}_\sigma^m$ be the m-homogeneous polynomial in $X_{j_1}, X_{j_2}, \cdots, X_{j_r}, Z_j (j \ne j_1, j_2, \cdots, j_r)$, given by setting $x_{j_1 0} = x_{j_2 0} = \cdots = x_{j_r 0} = 0$ in $\tilde{P}^m$.

In this way, for P, Q, H in Section 2, we can define $\tilde{P}^m, \tilde{Q}^m, \tilde{H}^m$ and $\tilde{P}_\sigma^m, \tilde{Q}_\sigma^m, \tilde{H}_\sigma^m$.

Proposition 3.2 For Q in Section 2, under the assumption in Theorem 2.1, $\tilde{Q}_\sigma^m$ has only trivial zeros (i.e. at least one of $X_{j_1}, X_{j_2}, \cdots, X_{j_r}, Z_j (j \ne j_1, j_2, \cdots, j_r)$ is zero) for any set $\sigma = (j_1, j_2, \cdots, j_r)(r \ge 1)$.

Proof. Let $X_{j_1}, X_{j_2}, \cdots, X_{j_r}, Z_j (j \ne j_1, j_2, \cdots, j_r)$ be a zero of $\tilde{Q}_\sigma^m$. Without loss of generality, let $x_{j0} = 1 (j \ne j_1, j_2, \cdots, j_r)$. Then

$$A_j X_j = 0, \quad j = j_1, j_2, \cdots, j_r$$
$$A_j X_j = B_j, \quad j \ne j_1, j_2, \cdots, j_r$$

where $A_j(r_j \times k_j), B_j(r_j \times 1)$ are given by the coefficients of $Q(X)$, and $\sum_{j=1}^m r_j = n$.

By the assumption in Theorem 2.1, A_j has no more than k_j rows for $j \ne j_1, j_2, \cdots, j_r$, so that there must be at least one s such that A_{j_s} has more than k_{j_s} rows. Thus $X_{j_s} = 0$. That is to say, $X_{j_1}, X_{j_2}, \cdots, X_{j_r}, Z_j (j \ne j_1, j_2, \cdots, j_r)$ is a trivial zero of $\tilde{Q}_\sigma^m$.

Lemma 3.3. Under the assumption in theorem 2.1, there exists a finite subset $S_1 \subset C$ such that, if $s \notin S_1, \tilde{Q}_\sigma^m + s\tilde{P}_\sigma^m$ has only trivial zeros for any set σ.

Proof. By Lemma 3.1, for any set σ, there exists a set of polynomials: $g_1(s), g_2(s), \cdots, g_t(s)$ such that for a given $s \in C, \tilde{Q}_\sigma^m + s\tilde{P}_\sigma^m$ has a nontrivial zero if and only if $g_1(s) = g_2(s) = \cdots = g_t(s) = 0$. We know that $g_1, g_2, \cdots, g_t$ are not all identically-zero polynomials because $\tilde{Q}_\sigma^m$ has only trivial zero. Let S_σ be the set of common zeros of $g_1, g_2, \cdots, g_t$ and $S_1 = \bigcup_\sigma S_\sigma$. Then for $s \notin S_1, \tilde{Q}_\sigma^m + s\tilde{P}_\sigma^m$ has only trivial zeros for any set σ.

Lemma 3.4. Under the assumption in Theorem 2.1, there exists a finite subset $S_2 \subset C$ such that 0 is a regular value of $Q(X) + sP(X)$ for $s \notin S_2$.

Lemma 3.4 follows from Proposition 3.5 below and the standard discussions in Zulehner [27] (also see the appendix in Li and Sauer [16]).

Proposition 3.5. Under the assumption in Theorem 2.1, $Q(X)$ has D_m nonsingular zeros.

This can be checked by direct computation because of the simple form of $Q(X)$.

Proof of Theorem 2.1. Let S be the set of principal values in $[0,2\pi)$ of negative angles of the complex numbers in $S_1 \cup S_2$. If $\theta \notin S$, then $t/(1-t)c \notin S_1 \cup S_2$ for $c = re^{i\theta} (r \ne 0$ real) and $t \ne 1$. By Lemma 3.4, 0 is a regular value of H. According to the discussions in Chow, Mallet-Paret and Yorke [4], we need only to show that, for $t_0 \in [0,1)$, there exists no curve that runs away to infinity when $t \longrightarrow t_0$. If this is not true, there exists a sequence $\{(X(t_k),t_k)\}_{k=1}^{\infty}, H(X(t_k),t_k)=0$, such that $t_k \longrightarrow t_0, ||X(t_k)|| \longrightarrow \infty$ when $k \longrightarrow \infty$. We suppose that $||X_{j_1}(t_k)|| \longrightarrow \infty, ||X_{j_2}(t_k)|| \longrightarrow \infty, \cdots, ||X_{j_r}(t_k)|| \longrightarrow \infty$ when $k \longrightarrow \infty$ and $||X_j(t_k)||(j \neq j_1, j_2, \cdots, j_r)$ are bounded. Let H_σ^m be given by setting $x_{j0} = 1(j \neq j_1, j_2, \cdots, j_r)$ in $\bar{H}_\sigma^m, H_\sigma^m = (h_{1\sigma}^m, h_{2\sigma}^m, \cdots, h_{n\sigma}^m)$. Apparently $h_{i\sigma}^m$ is the highest r-homogeneous part of h_i with respect to $X_{j_1}, X_{j_2}, \cdots, X_{j_r}$, and

$$\begin{aligned} & h_{i\sigma}^m(\cdots, X_{j_1}/||X_{j_1}||, \cdots, X_{j_2}/||X_{j_2}||, \cdots, X_{j_r}/||X_{j_r}||, \cdots) \\ & = ||X_{j_1}||^{-d_{ij_1}} \cdots ||X_{j_r}||^{-d_{ij_r}} h_{i\sigma}^m(\cdots, X_{j_1}, \cdots, X_{j_2}, \cdots, X_{j_r}, \cdots) \\ & = ||X_{j_1}||^{-d_{ij_1}} \cdots ||X_{j_r}||^{-d_{ij_r}} (h_i(X) - r_i(X)) \end{aligned} \tag{3.1}$$

where $r_i(X)$ consists of all terms of h_i whose degree with respect to X_{j_s}, for at lease one $s, 1 \le s \le r$, is lower than d_{ij_s}.

Because of the boundedness of $X_j(t_k)/||X_j(t_k)||$ $(j = j_1, j_2, \cdots, j_r)$ and $X_j(t_k)(j \neq j_1, j_2, \cdots, j_r)$, there is a subsequence of

$$\{(\cdots, X_{j_1}(t_k)/||X_{j_1}(t_k)||, \cdots, X_{j_2}(t_k)/||X_{j_2}(t_k)||, \cdots, X_{j_r}(t_k)/||X_{j_r}(t_k)||, \cdots)\}$$

converging to $X^*(t_0)$.

From

$$H(X(t_k), t_k) = 0,$$

$$||X_{j_1}||^{-d_{ij_1}} \cdots ||X_{j_r}||^{-d_{ij_r}} ||r_i(X(t_k))|| \longrightarrow 0, \quad k \longrightarrow \infty$$

and (3.1), we have

$$h_{i\sigma}^m(X^*(t_0), t_0) = 0, \quad i = 1, 2, \cdots, n$$

and

$$X^*(t_0) \neq 0$$

that is, H_σ^m has one non-trivial zero at t_0, which contradicts Lemma 3.3.

4. Discussions about $Q(X)$

In some cases, the start system $Q(X)$ can be constructed by hand. For general system, if it is difficult to construct $Q(X)$ by hand, we can construct it by choosing a_{jk}^i, b_{jl}^i randomly. This is the reason why we call it random product homotopy.

The following proposition ensures that the method works with probability one.

Proposition 4.1. For $1 \le j \le m$, there exists an open dense subset $U_i \subset C^{nk_j}$ such that if $(a_{j1}^1, \cdots, a_{jk_j}^1, \cdots, a_{j1}^n, \cdots, a_{jk_j}^n) \in U_1$, then $A(I_j)$ is nonsingular for any set I_j. There exists an open dense subset $U_2 \subset C^{d_{1j}+d_{2j}+\cdots+d_{nj}}$ such that if $(a_{j1}^1, \cdots, a_{jk_j}^1, \cdots, a_{j1}^n, \cdots, a_{jk_j}^n) \in U_1, (b_{j1}^1, b_{j2}^1, \cdots, \cdots, b_{j1}^n, b_{j2}^n, \cdots) \in U_2$, then the linear systems

$$A(I_j)X_j = B(I_j, L_j)$$

and

$$A(I'_j)X_j = B(I'_j, L'_j)$$

have different solutions for any sets $(I_j, L_j) \neq (I'_j, L'_j)$.

Proof. For a set I_j, $\det(A(I_j))$ is a homogeneous polynomial in $\{a^i_{jk}\}$, not identically zero. The union of zero sets of $\det(A(I_j))$ for all sets I_j is a proper algebraic subset in affine space C^{nk_j} (cf. Hartshorne [9]), and its complement U_1 is open dense in C^{nk_j}. If $(a^1_{j1}, \cdots, a^1_{jk_j}, \cdots, a^n_{j1}, \cdots, a^n_{jk_j}) \in U_1$, then $A(I_j)$ is nonsingular for any set I_j. The rest of the proposition can be proved similarly.

For randomly chosen a^i_{jk}, b^i_{jl}, if they do not satisfy the condition of Theorem 2.1 we can change parts of them randomly to satisfy the condition.

5. An application

Our method has two applications.

The first one is the general λ-matrix problem (cf. Lancaster [10])

$$\begin{aligned} &(A_n\lambda^n + A_{n-1}\lambda^{n-1} + \cdots + A_0)x = 0 \\ &C^{\top}x = 1 \end{aligned} \tag{5.1}$$

The second one is the multiparameter eigenvalue problem (cf. Atkinson [2])

$$\begin{aligned} &(A_{10} + \lambda_1 A_{11} + \cdots + \lambda_k A_{1k})x_1 = 0 \\ &\cdots \\ &(A_{k0} + \lambda_1 A_{k1} + \cdots + \lambda_k A_{kk})x_k = 0 \\ &C_1^{\top}x_1 = 1 \\ &\cdots \\ &C_k^{\top}x_k = 1 \end{aligned} \tag{5.2}$$

A homotopy method for problem (5.1) has been presented by Chu, Li and Sauer [5]. (5.1) and (5.2) can be put in the same pattern, the nonlinear multiparameter eigenvalue problem

$$\begin{aligned} &A_1(\lambda_1, \cdots, \lambda_k)x_1 = 0 \\ &\cdots \\ &A_k(\lambda_1, cdots, \lambda_k)x_k = 0 \\ &C_1^{\top}x_1 = 1 \\ &\cdots \\ &C_k^{\top}x_k = 1 \end{aligned} \tag{5.3}$$

where $A_i (i = 1, 2, \cdots, k)$ are $m_i \times m_i$ matrices whose elements are polynomials in $\lambda_1, \lambda_2, \cdots, \lambda_k$ of degree d_i respectively, and $x_i = (x_{i1}, x_{i2}, \cdots, x_{im_i})$.

When $k=1$, (5.3) degenerates into (5.1).

When $d_i = 1(i = 1,2,\cdots,k)$, (5.3) degenerates into (5.2).

When $k=1, d_i = 1$, (5.3) degenerates into the ordinary algebraic eigenvalue problem (1.2).

The variables can be naturally put into $k+1$ sets, $X_0 = (\lambda_1, \lambda_2, \cdots, \lambda_k), X_i = x_i(i = 1,2,\cdots,k)$. Then the $(k+1)$-Bezout number D_{k+1} of (5.3) is the coefficient of $t_0^k t_1^{m_1} \cdots t_k^{m_k}$ in

$$(d_1 t_0 + t_1)^{m_1} \cdots (d_k t_0 + t_k)^{m_k} t_1 \cdots t_k$$

that is

$$D_{k+1} = (d_1 m_1) \cdots (d_k m_k) = \prod_{i=1}^{k} d_i m_i$$

To solve (5.3), we construct the homotopy

$$H(X,t) = (1-t)cQ(X) + tP(X) \tag{5.4}$$

where $X = (X_0, X_1, \cdots, X_k), P(X)$ is the system in (5.3), and

$$Q = (q_{11}, \cdots, q_{1m_1}, \cdots, q_{k1}, \cdots, q_{km_k}, q_1, \cdots, q_k)$$

is as follows:

$$\begin{aligned} q_{ij}(X) &= (\lambda_i^{d_i} - a_{ij})x_{ij} \quad i = 1,2,\cdots,k, j = 1,2,\cdots,m_i \\ q_i(X) &= x_{i1} + \cdots + x_{im_i} - 1, \quad i = 1,2,\cdots,k \end{aligned} \tag{5.5}$$

$\lambda_i^{d_i} - a_{ij}$ can be factored into $\prod_{l=1}^{d_i}(\lambda_i - a_l^{ij}), a_l^{ij}, l = 1,2,\cdots,d_i$ are d_i-roots of a_{ij}.

It is easy to see that $Q(X)$ in (5.5) satisfies the condition of Theorem 2.1, if

$$(a_{ij})^s \neq a_{ij'}, (j \neq j')$$

for any integer s.

Notation. Having finished this paper, we read a paper [25] of Verschelde, Beckers and Haegemans, in which a different idea to construct the start system for solving polynomial systems in $C^{k_1} \times C^{k_2} \times \cdots \times C^{k_m}$ is presented. But their start system can not serve all such systems, and their start system needs computing D_m determinants of Jacobian matrices to check whether the start system has exactly D_m nonsingular solutions.

References

[1] E.Allgower and K.Georg, "Numerical Continuation Methods", Springer-Verlag, New York, 1990.

[2] F.V.Atkinson, "Multiparameter Eigenvalue Problem", Vol.1 Marices and Compact Operators, Academic Press, New York and London, 1972.

[3] P.Brunovsky & P.Meravy, "Solving Systems of Polynomial Equations by Bounded and Real Homotopy", Numer. Math. 43 (1984), 397-418.

[4] S.N.Chow, J.Mallet-Paret and J.A.Yorke, "Homotopy Method for Locating All Zeros of a System of Polynomials", In Lecture Notes in Mathematics No. 730 1979, 77-88.

[5] M.Chu, T.Y.Li and T.Sauer, "A Homotopy Method for General λ-matrix Problems", SIAM J.Matrix. Anal., 9 (1988), 528-536.

[6] F.J.Drexler, "Eine Methode zur Berechnung Samtlicher Losungen von Polynomgleichungssystemen", Numer. Math. 29 (1977), 45-58.

[7] C.B.Garcia and T.Y.Li, "On the Number of Solutions to Polynomial Systems of Equations", SIAM J.Numer. Anal. 17 (1980), 540-546.

[8] C.B.Garcia and W.I.Zangwill, "Finding All Solutions to Polynomial Systems and Other Systems of Equations", Math. Programming 16 (1979), 159-176.

[9] R.Hartshorne, "Algebraic Geometry", GTM, 52, Springer-Verlag, New York, 1997.

[10] P.Lancaster, "Lambda-matrices and Vibrating Systems", Pergmon, Oxford, 1966.

[11] T.Y.Li, "On Chow, Mallet-Paret and Yorke Homotopy for Solving Systems of Polynomials", Bull. Inst. Math. Acad. Science 11 (1983), 433-437.

[12] T.Y.Li, T.Sauer and J.Yorke, "Numerical Solution of A Class of Deficient Polynomial Systems", SIAM J.Numer. Anal. Vol.24, No.2 (1987). 435-451.

[13] T.Y.Li, T.Sauer and J.Yorke, "The Random Product Homotopy and Deficient Polynomial Systems", Numer. Math. 51, (1987), 481-500.

[14] T.Y.Li, T.Sauer and J.Yorke, "The Cheater's Homotopy: an Efficient Procedure for Solving Systems of Polynomial Equations", SIAM J.Numer. Anal., Vol.26, No.5 (1989), 1241-1251.

[15] T.Y.Li, T.Sauer and J.Yorke, "Numerical Determining Solutions of Systems of Polynomial Equations", Bull. A. M. S. 18 (1988), 173-177.

[16] T.Y.Li and T.Sauer, "A Simple Homotopy for Solving Deficient Polynomial Systems", Japan J.Appl. Math. 6 (1989), 409-419.

[17] T.Y.Li and X.Wang, "Solving Deficient Polynomial System with Homotopies Which Keep the Subschemes at Infinity Invariant", Math. Comp. Vol.56, No.194

(1991), 693-710.

[18] A.Morgan, "A Transformation to Avoid Solutions at Infinity for Polynomial Systems", Appl. Math. Comput. 18 (1986), 77-86.

[19] A.Morgan, "A Homotopy for Solving Polynomial Systems", Appl. Math. Comput. 18 (1986), 87-92.

[20] A.Morgan and A.Sommese, "A Homotopy for Solving General Polynomial Systems That Respect m-homogeneous Structures", Appl. Math. Comput. 24 (1987), 95-114.

[21] A.Morgan and A.Sommese, "Computing All Solutions to Polynomial System Using Homotopy Continuation", Appl. Math. Comput. 24 (1987), 115-138.

[22] G.L.Naber, "Topological Methods in Euclidean Spaces", Cambrige Univ. Press. London, 1990.

[23] I.R.Shafarevich, "Basic Algebraic Geometry", Springer-Verlag, New York, 1977.

[24] B.L.Van Der Waerden, "Modern Algebra", Vol.2, Ungar, New York, 1951.

[25] J.Verschelde, M.Beckers and A.Haegemans, "A New Start System for Solving Deficient Polynomial Systems Using Continuation", Appl. Math. Comput. 44 (1991), 225-239.

[26] A.H.Wright, "Finding All Solutions to System of Polynomial Equations", Math. Comp. Vol.44 (1985), 125-133.

[27] W.Zulehner, "Homotopy Method for Determining All Isolated Solutions of Polynomial Systems", Math. Comp. 50 (1988), 167-177.

Computer Mathematics
Proc. of the Special Program at
Nankai Institute of Mathematics
January 1991 - June 1991

Parallel Computation Simulating Sequential Computation

Hu Guoding
Nankai Institute of Mathematics
Nankai University, Tianjin, P.R.China

1 Introduction

We have established in [1] a general form of sequential computation with a single computing machine. On the basis of [1], we shall design a mechanical program in this paper, for the simulation of any given sequential computation with a single "uniform separable" (US) computing machine, by a corresponding parallel computation with several computing machines.

We shall show that for any computation process P

$$\alpha^{(0)} \to \alpha^{(1)} \to \alpha^{(2)} \to \cdots \to \alpha^{(n)}$$

of a given US computing machine M , we can always select its corresponding outline process

$$\alpha^{(k_0)} \to \alpha^{(k_1)} \to \cdots \to \alpha^{(k_m)} \quad (0 = k_0 < k_1 < \cdots < k_m = n)$$

and the classification $I_l (l = 0, 1, 2, \cdots, N)$ of set

$$\{0, 1, 2, \cdots, m-1\}$$

such that the serial P can be simulated by the parallel computations

$$\alpha^{(k_i)} \longrightarrow \alpha^{(k_{i+1})} \quad (i \in I_l)$$

successively as $l = 0, 1, 2, \cdots, N$.

The "uniformly separable" condition for the computing machine seems not only sufficient but also necessary to the general possibility for designing the mechanical program mentioned above.

In [2], a static definition of US computing machines is given, and two important results are proved (All total recursive functions in a one-symbol-coded system and all primitive recursive functions in a binary system can be computed by a US computing machine.) In [3], according to the method given in this paper, a concrete parallel program (written in data flow parallel language PII, see [4]) is constructed.

2 Uniformly separable (US) mapping

Let us consider the mapping $\tilde{G}$,

$$\alpha_{\lambda,\mu} \xrightarrow{\tilde{G}} \alpha'_{\lambda',\mu'}$$

where (i) $\lambda, \mu(\lambda', \mu')$ are sequences of letter coordinates in $\alpha(\alpha')$, and $\alpha_{\lambda,\mu}(\alpha'_{\lambda',\mu'})$ is word $\alpha(\alpha')$ with arrows above or below those letters whose coordinates are in $\lambda(\lambda')$ or $\mu(\mu')$ respectively; (ii) the mapping $\tilde{G}$ is independent of μ.

1[0] Definition of the simple FC mapping $\tilde{G}$.

Here we restrict ourselves to considering $\lambda, \mu(\lambda', \mu')$ in $\tilde{G}$ as bounded numbers of letter coordinates in $\alpha(\alpha')$.

Following 6[0], §2 of [1], the simple FC mapping $\tilde{G}$ must be determined by

$$\alpha_{\lambda,\mu} \xrightarrow{\widetilde{K}} \tilde{\alpha} \xrightarrow{\tilde{\tilde{G}}} \widetilde{\alpha'} \xrightarrow{\tilde{J}} \alpha'_{\lambda',\mu'}$$

in which (1) $\widetilde{K}$ and $\tilde{J}$ are just the same as in 6[0], §2 of [1]; and (2) $\tilde{\tilde{G}}$ is a mapping on words with bounded length:

$$\tilde{\alpha} \in E \xrightarrow{\tilde{\tilde{G}}} \widetilde{\alpha'} \in E'$$

where

$$E = A \cup V (A \cap V = \Phi), V \text{ is finite},$$

$$E' = \cup_{i=1}^{8} E'_i, E'_1 = A, E'_2 = \overset{\downarrow}{A} = \{\overset{\downarrow}{a} | a \in A\},$$

$$E'_3 = \underset{\uparrow}{A} = \{\underset{\uparrow}{a} | a \in A\}, E'_4 = \underset{\uparrow}{\overset{\downarrow}{A}} = \{\underset{\uparrow}{\overset{\downarrow}{a}} | a \in A\},$$

$$E'_5 = V, E'_6 = {}^{\downarrow}V = \{{}^{\downarrow}v | v \in V\},$$

$$E'_7 = V^{\downarrow} = \{v^{\downarrow} | v \in V\}, E'_8 = {}^{\downarrow}V^{\downarrow} = \{{}^{\downarrow}v^{\downarrow} | v \in V\}.$$

2[0]. Definition and properties of the US mapping $\tilde{G}$.

(2.1) The mapping $\tilde{G}$:

$$\alpha_{\lambda,\mu} \xrightarrow{\tilde{G}} \alpha'_{\lambda',\mu'}$$

is said to be US, if it satisfies conditions (1) and (2) below.

(1) Uniformity.

To every word $\alpha = a_1 \cdots a_n$ with common length n and $\lambda = \{i_1, \cdots, i_k\}$, $\mu = \{j_1, \cdots, j_l\}$ $(1 \leq i_1 < \cdots < i_k \leq n,\ 1 \leq j_1 < \cdots < j_l \leq n)$, there corresponds a word $\alpha' = a'_1 \cdots a'_{n'}$ with common length n' and $\lambda' = \{i'_1, \cdots, i'_{k'}\}$, $\mu' = \{j'_1, \cdots, j'_{l'}\}$ $(1 \leq i'_1 < \cdots < i'_{k'} \leq n', 1 \leq j'_1 < \cdots < j'_{l'} \leq n')$.

(2) Separation.

To every word $\alpha = a_1 \cdots a_n$ with common n and λ, the corresponding word $\alpha' = a'_1 \cdots a'_{n'}$ with common n' and μ', may be deducted in parallel by two separate parts:

(i) there exists a mapping (G) such that

$$(a_{i_1}, \cdots, a_{i_k}) \overset{(G)}{\longrightarrow} (a'_{j'_1}, \cdots, a'_{j'_{l'}})$$

and there is no $a'_{j'}(j \in \mu')$ which is identically equal to a certain a_i;

(ii) each letter occurence $a'_{j'}(j' \notin \mu', 1 \leq j' \leq n')$ is identically equal to a certain $a_i (1 \leq i \leq n)$.

(2.2) For any US mapping G, the letter occurrence $a_i (i \in \lambda)$ is called G-operating; another letter occurrence $a_i (i \notin \lambda)$ is called G-nonoperating. The letter occurrence $a'_{j'}(j' \in \mu')$ is called G-newborn; another letter occurence $a'_{j'}(j' \notin \mu')$ is called G-reserved.

We use

$$S(\alpha), T(\alpha')$$
$$(\overline{S}(\alpha), \overline{T}(\alpha'))$$

to denote the sequences of G-operating letter occurrences and G-newborn letter occurrences (G-nonoperating letter occurrences and G-reserved letter occurrences) respectively. Considering that the mapping (G) is uniquely determined by G, the separability condition of US mapping G may be simply expressed as follows:

(i) $S(\alpha) \overset{G}{\longrightarrow} T(\alpha')$

(ii) $\overline{T}(\alpha')$ is equal to a certain sequence of letter occurrences in α.

Let us use an example to illustrate a US and simple FC mapping $\tilde{G}$ by a particular correspondence:

$$\alpha_{\lambda,\mu} \rightarrow \tilde{\alpha} \overset{\tilde{\tilde{G}}}{\longrightarrow} \widetilde{\alpha'} \rightarrow \alpha'_{\lambda',\mu'}$$

in which related words are as follows:

$$\alpha = a_1 a_2 \cdots a_{11}$$

$$\begin{aligned}
\alpha_{\lambda,\mu} &= a_1\overset{\downarrow}{a_2}\underset{\uparrow}{a_3}a_4\overset{\downarrow}{a_5}a_6a_7a_8\underset{\uparrow}{\overset{\downarrow}{a_9}}a_{10}a_{11}\\
\lambda &= \{2,5,9\}, \mu = \{3,9\}\\
\tilde\alpha &= v_1a_1v_2a_5v_3a_9v_4\\
v_1 &= a_1, v_2 = a_3a_4, v_3 = a_6a_7a_8, v_4 = a_{10}a_{11}\\
\widetilde{\alpha'} &= \tilde{\tilde G}(\tilde\alpha) = v_2c_1a_5v_4c_2v_3a_2c_3c_4
\end{aligned}$$

$$\begin{aligned}
\alpha'_{\lambda',\mu'} &= \tilde J(\widetilde{\alpha'})\\
&= a_3\overset{\downarrow}{a_4}\underset{\uparrow}{c_1}\overset{\downarrow}{a_5}a_{10}a_{11}\underset{\uparrow}{c_2}\overset{\downarrow}{a_6}a_7\overset{\downarrow}{a_8}a_2\underset{\uparrow}{\overset{\downarrow}{c_3}}\underset{\uparrow}{c_4}\\
&= a'_1\overset{\downarrow}{a'_2}\underset{\uparrow}{a'_3}\overset{\downarrow}{a'_4}a'_5a'_6\underset{\uparrow}{a'_7}\overset{\downarrow}{a'_8}a'_9\overset{\downarrow}{a'_{10}}a'_{11}\underset{\uparrow}{\overset{\downarrow}{a'_{12}}}\underset{\uparrow}{a'_{13}}\\
a'_1 &= a_3, a'_2 = a_4, a'_3 = c_1, a'_4 = a_5, a'_5 = a_{10}, a'_6 = a_{11},\\
a'_7 &= c_2, a'_8 = a_6, a'_9 = a_7, a'_{10} = a_8, a'_{11} = a_2, a'_{12} = c_3, a'_{13} = c_4\\
\lambda' &= \{2,4,8,10,12\}, \mu' = \{3,7,12,13\}
\end{aligned}$$

It is easy to see that the above correspondence of US mapping $\tilde G$,

$$\alpha = a_1a_2\cdots a_{11} \overset{\tilde G}{\longrightarrow} \alpha' = a'_1a'_2\cdots a'_{13},$$

may always be deducted in parallel by two separate parts:

(i)

$$\begin{aligned}
S(\alpha) = (a_2, a_5, a_9) \overset{\tilde G}{\longrightarrow} T(\alpha') &= (a'_3, a'_7, a'_{12}, a'_{13})\\
&= (c_1, c_2, c_3, c_4)
\end{aligned}$$

(ii)

$$\begin{aligned}
\overline{T}(\alpha') &= (a'_1, a'_2, a'_4, a'_5, a'_6, a'_8, a'_9, a'_{10}, a'_{11})\\
&= (a_3, a_4, a_5, a_{10}, a_{11}a_6, a_7, a_8, a_2).
\end{aligned}$$

3^0 <u>Composite mapping of several US mappings.</u>

Let $G = G_{q^{(n-1)}}G_{q^{(n-2)}}\cdots G_{q^{(0)}}$ be the composite mapping of mappings $G_{q^{(k)}}$ $(k = 0,1,\cdots,n-1)$,

$$\begin{aligned}
G(\alpha) = G(\alpha^{(0)}) &= G_{q^{(n-1)}}\cdots G_{q^{(0)}}(\alpha^{(0)})\\
&= G_{q^{(n-1)}}\cdots(\alpha^{(1)})\\
&= \cdots\\
&= G_{q^{(n-1)}}(\alpha^{(n-1)}) = \alpha^{(n)} = \alpha'
\end{aligned}$$

If the mappings $G_{q^{(k)}}(k = 0, 1, 2, \cdots, n-1)$ are all US, then G is also US and

(i) letter occurrence of α is G-operating iff there exists $0 \le k < n-1$ such that this letter occurrence is $G_{q^{(k)}}$-operating. Otherwise, letter occurrence of α is $G_{q^{(k)}}$-nonoperating.

(ii) letter occurrence of α' is G-reserved iff to every $0 \le k < n-1$, this letter is $G_{q^{(k)}}$-reserved. Otherwise, letter occurrence of α' is $G_{q^{(k)}}$-newborn.

Now we illustrate it by a particular example. Let US and FC composite mapping $\tilde{G} = \tilde{G}_{q^{(1)}}\tilde{G}_{q^{(0)}}$ of US and simple FC mappings $\widetilde{G_{q^{(k)}}}(k = 0, 1)$ be given, i.e.

$$\alpha = \alpha^{(0)} \rightarrow \alpha^{(0)}_{\lambda^{(0)},\mu^{(0)}} \xrightarrow{\tilde{G}_{q^{(0)}}} \alpha^{(1)}_{\lambda^{(1)},\mu^{(1)}} \xrightarrow{\tilde{G}_{q^{(1)}}} \alpha^{(2)}_{\lambda^{(2)},\mu^{(2)}} \rightarrow \alpha^{(2)} = \alpha'$$

in which each

$$\alpha^{(k)}_{\lambda^{(k)},\mu^{(k)}} \xrightarrow{\tilde{G}_{q^{(k)}}} \alpha^{(k+1)}_{\lambda^{(k+1)},\mu^{(k+1)}}$$

is determined by

$$\alpha^{(k)}_{\lambda^{(k)},\mu^{(k)}} \xrightarrow{\widetilde{K}^{(k)}} \tilde{\alpha}^{(k)} \xrightarrow{\tilde{\tilde{G}}_{q^{(k)}}} \widetilde{\alpha'}^{(k)} \xrightarrow{\tilde{J}^{(k+1)}} \alpha^{(k+1)}_{\lambda^{(k+1)},\mu^{(k+1)}} (k = 0, 1)$$

where

$$\alpha = \alpha^{(0)} = a_1^{(0)} a_2^{(0)} \cdots a_{11}^{(0)}$$

$$\alpha^{(0)}_{\lambda^{(0)},\mu^{(0)}} = a_1^{(0)} \overset{\downarrow}{a}_2^{(0)} \underset{\uparrow}{a}_3^{(0)} a_4^{(0)} \overset{\downarrow}{a}_5^{(0)} a_6^{(0)} a_7^{(0)} a_8^{(0)} \underset{\uparrow}{\overset{\downarrow}{a}}_9^{(0)} a_{10}^{(0)} a_{11}^{(0)}$$

$$\lambda^{(0)} = \{2, 5, 9\}, \mu^{(0)} = \{3, 9\}$$

$$\tilde{\alpha}^{(0)} = v_1^{(0)} a_2^{(0)} v_2^{(0)} a_5^{(0)} v_3^{(0)} a_9^{(0)} v_4^{(0)}$$

$$v_1^{(0)} = a_1^{(0)}, v_2^{(0)} = a_3^{(0)} a_4^{(0)}, v_3^{(0)} = a_6^{(0)} a_7^{(0)} a_8^{(0)}, v_4^{(0)} = a_{10}^{(0)} a_{11}^{(0)}$$

$$\widetilde{\alpha'}^{(0)} = \tilde{\tilde{G}}_{q^{(0)}}(\tilde{\alpha}^{(0)}) = \overset{\downarrow}{v}_2^{(0)} \underset{\uparrow}{c}_1^{(1)} \overset{\downarrow}{a}_5^{(0)} v_4^{(0)} \underset{\uparrow}{c}_2^{(0)} \overset{\downarrow\downarrow}{v}_3^{(0)} a_2^{(0)} \underset{\uparrow}{\overset{\downarrow}{c}}_3^{(1)} \underset{\uparrow}{c}_4^{(1)}$$

$$\begin{aligned} \alpha^{(1)}_{\lambda^{(1)},\mu^{(1)}} &= a_3^{(0)} \overset{\downarrow}{a}_4^{(0)} \underset{\uparrow}{c}_1^{(1)} \overset{\downarrow}{a}_5^{(0)} a_{10}^{(0)} a_{11}^{(0)} \underset{\uparrow}{c}_2^{(1)} \overset{\downarrow}{a}_6^{(0)} a_7^{(0)} \overset{\downarrow}{a}_8^{(0)} a_2^{(0)} \underset{\uparrow}{\overset{\downarrow}{c}}_3^{(1)} \underset{\uparrow}{c}_4^{(1)} \\ &= a_1^{(1)} \overset{\downarrow}{a}_2^{(1)} \underset{\uparrow}{a}_3^{(1)} \overset{\downarrow}{a}_4^{(1)} a_5^{(1)} a_6^{(1)} \underset{\uparrow}{a}_7^{(1)} \overset{\downarrow}{a}_8^{(1)} a_9^{(1)} \overset{\downarrow}{a}_{10}^{(1)} a_{11}^{(1)} \underset{\uparrow}{\overset{\downarrow}{a}}_{12}^{(1)} a_{13}^{(1)} \end{aligned}$$

$$a_1^{(1)} = a_3^{(0)}, a_2^{(1)} = a_4^{(0)}, a_3^{(1)} = c_1^{(1)}, a_4^{(1)} = a_5^{(0)}, a_5^{(1)} = a_{10}^{(0)}, a_6^{(1)} = a_{11}^{(0)},$$
$$a_7^{(1)} = c_2^{(1)}, a_8^{(1)} = a_6^{(0)}, a_9^{(1)} = a_7^{(0)}, a_{10}^{(1)} = a_8^{(0)}, a_{11}^{(1)} = a_2^{(0)}, a_{12}^{(1)} = c_3^{(1)},$$
$$a_{13}^{(1)} = c_4^{(1)}$$

$\lambda^{(1)} = \{2,4,8,10,12\}, \mu^{(1)} = \{3,7,12,13\}$

$$\widetilde{\alpha^{(1)}} = v_1^{(1)} a_2^{(1)} v_2^{(1)} a_4^{(1)} v_3^{(1)} a_8^{(1)} v_4^{(1)} a_{10}^{(1)} v_5^{(1)} a_{12}^{(1)} v_6^{(1)}$$

$v_1^{(1)} = a_1^{(1)} = a_3^{(0)}, v_2^{(1)} = a_3^{(1)} = c_1^{(1)}, v_3^{(1)} = a_5^{(1)} a_6^{(1)} a_7^{(1)} = a_{10}^{(0)} a_{11}^{(0)} c_2^{(1)},$
$v_4^{(1)} = a_9^{(1)} = a_7^{(0)}, v_5^{(1)} = a_{11}^{(1)} = a_2^{(0)}, v_6^{(1)} = a_{13}^{(1)} = c_4^{(1)}$

$$\widetilde{\alpha'^{(1)}} = \tilde{\tilde{G}}_{q^{(1)}}(\tilde{\alpha}^{(1)}) = \overset{\downarrow\downarrow}{v}{}_3^{(1)} \underset{\uparrow}{c}{}_1^{(2)} a_4^{(1)} \overset{\downarrow}{v}{}_5^{(1)} \overset{\downarrow}{a}{}_{12}^{(1)} a_8^{(1)} v_6^{(1)} \underset{\uparrow}{\overset{\downarrow}{c}}{}_2^{(2)}$$

$$\begin{aligned}
\alpha^{(2)}_{\lambda^{(2)},\mu^{(2)}} &= \overset{\downarrow}{a}{}_{10}^{(0)} a_{11}^{(0)} \overset{\downarrow}{c}{}_2^{(1)} \underset{\uparrow}{c}{}_1^{(2)} a_4^{(1)} \overset{\downarrow}{a}{}_2^{(0)} \overset{\downarrow}{a}{}_{12}^{(1)} a_8^{(1)} c_4^{(1)} \underset{\uparrow}{\overset{\downarrow}{c}}{}_2^{(2)} \\
&= \overset{\downarrow}{a}{}_{10}^{(0)} a_{11}^{(0)} \overset{\downarrow}{c}{}_2^{(1)} \underset{\uparrow}{c}{}_1^{(2)} a_5^{(0)} \overset{\downarrow}{a}{}_2^{(0)} \overset{\downarrow}{c}{}_3^{(1)} a_6^{(0)} c_4^{(1)} \underset{\uparrow}{\overset{\downarrow}{c}}{}_2^{(2)} \\
&= \overset{\downarrow}{a}{}_1^{(2)} a_2^{(2)} \overset{\downarrow}{a}{}_3^{(2)} \underset{\uparrow}{a}{}_4^{(2)} a_5^{(2)} \overset{\downarrow}{a}{}_6^{(2)} \overset{\downarrow}{a}{}_7^{(2)} a_8^{(2)} a_9^{(2)} \underset{\uparrow}{\overset{\downarrow}{a}}{}_{10}^{(2)}
\end{aligned}$$

$$a_1^{(2)} = a_{10}^{(0)}, a_2^{(2)} = a_{11}^{(0)}, a_3^{(2)} = c_2^{(1)}, a_4^{(2)} = c_1^{(2)}, a_5^{(2)} = a_5^{(0)},$$
$$a_6^{(2)} = a_2^{(0)}, a_7^{(2)} = c_3^{(1)}, a_8^{(2)} = a_6^{(0)}, a_9^{(2)} = c_4^{(1)}, a_{10}^{(2)} = c_2^{(2)}$$
$$\lambda^{(2)} = \{1,3,6,7,10\}, \mu^{(2)} = \{4,10\}.$$

We may find that G-operating letter occurrences of α are $a_2^{(0)}$, $a_5^{(0)}$, $a_6^{(0)}$, $a_8^{(0)}$, $a_9^{(0)}$, and G-reserved letter occurrences of α' are $a_1^{(2)} = a_{10}^{(0)}, a_2^{(2)} = a_{11}^{(0)}, a_5^{(2)} = a_5^{(0)}, a_6^{(2)} = a_2^{(0)}, a_8^{(2)} = a_6^{(0)}$.

3 US computing machine

1^0. First of all, we shall illustrate "multiplication" as a simple example of the US computing machine.

The computation process of "multiplication" with different words,

$$369 \times 234 \to 36\overset{\downarrow}{9}23\overset{\downarrow}{4} \to 3\overset{\downarrow}{6}923\overset{\downarrow}{4}\underset{\uparrow}{3}\underset{\uparrow}{6} \to \overset{\downarrow}{3}69234\overset{\downarrow}{3}... $$

$$369 \times 234 \to 36\overset{\downarrow}{9}23\overset{\downarrow}{4} \to 3\overset{\downarrow}{6}923\overset{\downarrow}{4}\underset{\uparrow}{3}\underset{\uparrow}{6} \to \overset{\downarrow}{3}6923\overset{\downarrow}{4}36\underset{\uparrow}{2}\underset{\uparrow}{4} \to \cdots$$

$$238 \times 025 \to 23\overset{\downarrow}{8}02\overset{\downarrow}{5} \to 2\overset{\downarrow}{3}802\overset{\downarrow}{5}\underset{\uparrow}{4}\underset{\uparrow}{0} \to \overset{\downarrow}{2}3802\overset{\downarrow}{5}40\underset{\uparrow}{1}\underset{\uparrow}{5} \to \cdots$$

may be summarized into a computation process of word as follows:

$$\begin{aligned}
& a_1^{(0)} a_2^{(0)} a_3^{(0)} \times a_4^{(0)} a_5^{(0)} a_6^{(0)} \\
\to \quad & a_1^{(0)} a_2^{(0)} \overset{\downarrow}{a}{}_3^{(0)} a_4^{(0)} a_5^{(0)} \overset{\downarrow}{a}{}_6^{(0)}
\end{aligned}$$

$$\rightarrow \quad a_1^{(0)}\overset{\downarrow}{a_2^{(0)}}a_3^{(0)}a_4^{(0)}a_5^{(0)}\overset{\downarrow}{a_6^{(0)}}\underset{\uparrow}{c_1^{(1)}}\underset{\uparrow}{c_2^{(1)}}$$

$$\rightarrow \quad \overset{\downarrow}{a_1^{(0)}}a_2^{(0)}a_3^{(0)}a_4^{(0)}a_5^{(0)}\overset{\downarrow}{a_6^{(0)}}c_1^{(1)}c_2^{(1)}\underset{\uparrow}{c_1^{(2)}}\underset{\uparrow}{c_2^{(2)}}$$

$$\rightarrow \quad \cdots\cdots$$

It is obvious that the first (second) step computation:

$$a_1^{(0)}a_2^{(0)}\overset{\downarrow}{a_3^{(0)}}a_4^{(0)}a_5^{(0)}\overset{\downarrow}{a_6^{(0)}} \rightarrow a_1^{(0)}\overset{\downarrow}{a_2^{(0)}}a_3^{(0)}a_4^{(0)}a_5^{(0)}\overset{\downarrow}{a_6^{(0)}}\underset{\uparrow}{c_1^{(1)}}\underset{\uparrow}{c_2^{(1)}}$$

$$(a_1^{(0)}\overset{\downarrow}{a_2^{(0)}}a_3^{(0)}a_4^{(0)}a_5^{(0)}\overset{\downarrow}{a_6^{(0)}}\underset{\uparrow}{c_1^{(1)}}\underset{\uparrow}{c_2^{(1)}} \rightarrow \overset{\downarrow}{a_1^{(0)}}a_2^{(0)}a_3^{(0)}a_4^{(0)}a_5^{(0)}\overset{\downarrow}{a_6^{(0)}}c_1^{(1)}c_2^{(1)}\underset{\uparrow}{c_1^{(2)}}\underset{\uparrow}{c_2^{(2)}})$$

of "multiplication" is a US and simple FC mapping.

2^0. Consider the mapping $\tilde{F}$:

$$(\alpha_{\lambda,\mu}, q) \overset{\tilde{F}}{\longrightarrow} (\alpha'_{\lambda',\mu'}, q')$$

where q, q' are state words with bounded length.

Following [1, §2, 7^0], the mapping $\tilde{F}$ will be US & simple FC iff

(i) q' dependent on q and m in $\tilde{\alpha}$ (cf. [1§2,6^0] only;

(ii) for each q, q', the mapping

$$\alpha_{\lambda,\mu} \longrightarrow \alpha'_{\lambda',\mu'}$$

produced by $\tilde{F}$ is US and simple FC as $\tilde{G}$ in 2^0 of §2.

3^0. A computing machine $M = \{A, W, Q, Q_h, \tilde{F}, P\}$ is said to be US if its computation rule $\tilde{F}$,

$$(\alpha_{\lambda,\mu}, q) \overset{\tilde{F}}{\longrightarrow} (\alpha'_{\lambda',\mu'}, q')$$

as a mapping described in 2^0, is US as well as simple FC.

In a US computing machine M, its computation process P

$$\alpha^{(0)}, \alpha^{(1)}, \cdots, \alpha^{(n)}$$

or

$$\alpha^{(0)}_{\lambda^{(0)},\mu^{(0)}}, \alpha^{(1)}_{\lambda^{(1)},\mu^{(1)}}, \cdots, \alpha^{(n)}_{\lambda^{(n)},\mu^{(n)}}$$

is obtained from

$$\alpha^{(0)}_{\lambda^{(0)},\mu^{(0)}} \overset{\tilde{G}_{q^{(0)}}}{\longrightarrow} \alpha^{(1)}_{\lambda^{(1)},\mu^{(1)}} \overset{\tilde{G}_{q^{(1)}}}{\longrightarrow} \cdots \overset{\tilde{G}_{q^{(n-1)}}}{\longrightarrow} \alpha^{(n)}_{\lambda^{(n)},\mu^{(n)}}$$

in which $\tilde{G}_{q^{(k)}}$ is produced by $\tilde{F}$,

$$(\alpha^{(k)}_{\lambda^{(k)},\mu^{(k)}}, q^{(k)}) \xrightarrow{\tilde{F}} (\alpha^{(k+1)}_{\lambda^{(k+1)},\mu^{(k+1)}}, q^{(k+1)})$$

where $k = 0, 1, 2, \cdots, n-1$.

Note that $\lambda^{(k)}(\mu^{(k+1)})$ is a coordinate sequence of $\tilde{G}_{q^{(k)}}$-operating letter occurrences ($\tilde{G}_{q^{(k)}}$-newborn letter occurrences) whenever $k = 0, 1, 2, \cdots, n-1$.

4⁰. Let M $= \{A, W, Q, Q_h, \tilde{F}, P\}$ be a US computing machine with typical computation process P:

$$\alpha^{(0)} \to \alpha^{(1)} \to \alpha^{(2)} \to \cdots \to \alpha^{(n)},$$

of composite mappings

$$\alpha^{(k)} \longrightarrow \alpha^{(k+1)} \quad k = 0, 1, 2, \cdots, n-1$$

The so-called computation outline process P′:

$$\alpha^{(0)} = \alpha^{(k_0)} \to \alpha^{(k_1)} \to \alpha^{(k_2)} \to \cdots \to \alpha^{(k_m)} = \alpha^{(n)}$$

$$0 = k_0 < k_1 < k_2 < \cdots < k_m = n$$

of P is constructed by composite mapping,

$$\alpha^{(k_i)} \longrightarrow \alpha^{k_{i+1})}, \quad 0 \le i \le m-1$$

each of which is the composite mapping of mappings

$$\alpha^{(k_v)} \longrightarrow \alpha^{(k_v+1)} \quad k_i \le k_v < k_{i+1}$$

and so is determined by

$$\alpha^{(k_i)} \to \alpha^{(k_i+1)} \to \cdots \to \alpha^{(k_{i+1})}$$

which is called a computation subprocess of P.

Thus, if we denote $(k_i|k_{i+1})$ as the mapping

$$\alpha^{(k_i)} \xrightarrow{(k_i|k_{i+1})} \alpha^{(k_{i+1})}$$

and $(k|k+1)$ as the mapping

$$\alpha^{(k)} \xrightarrow{(k|k+1)} \alpha^{(k+1)},$$

then the mapping $(k_i|k_{i+1})$ is the composite mapping of $(k|k+1), (k = k_i, k_i+1, \cdots, k_{i+1}-1)$, i.e.

$$(k_i|k_{i+1}) = (k_i|k_i+1)(k_i+1|k_i+2)\cdots(k_{i+1}-1|k_{i+1}) \quad (0 \le i \le m-1).$$

5^0. In a US computing machine M , we consider a computation outline process P':

$$\alpha^{(k_0)} \to \alpha^{(k_1)} \to \cdots \to \alpha^{(k_m)} \quad (0 = k_0 < k_1 < \cdots < k_m = n)$$

of P,

$$\alpha^{(0)} \to \alpha^{(1)} \to \alpha^{(2)} \to \cdots \to \alpha^{(n)}$$

Since each mapping $\alpha^{(k)} \xrightarrow{(k|k+1)} \alpha^{(k+1)}$ is US and so each $(k|k+1)$-operating (nonoperating) letter occurrences and $(k|k+1)$-newborn (reserved) letter occurrence have been given $(k = 0, 1, \cdots, n-1)$, each mapping

$$\alpha^{(k_i)} \xrightarrow{(k_i|k_{i+1})} \alpha^{(k_i+1)}$$

is also US and so each $(k_i|k_{i+1})$-operating (nonoperating) letter occurrences & $(k_i|k_i+1)$-newborn (reserved) letter occurrences will be calculated out from the above (see 3^0, §2).

6^0. Pure serial computation outline process (PSCOP)

(6.1) The PSCOP:

$$\alpha^{(k_0)} \to \alpha^{(k_1)} \to \alpha^{(k_2)} \to \cdots \to \alpha^{(k_m)} \quad (0 = k_0 < k_1 < \cdots < k_m = n)$$

of P:

$$\alpha^{(0)} \to \alpha^{(1)} \to \alpha^{(2)} \to \cdots \to \alpha^{(n)}$$

is defined as a computation outline process in which each

$$\alpha^{(k_i)} \xrightarrow{(k_i|k_{i+1})} \alpha^{(k_{i+1})}$$

is determined by the computation subprocess:

$$\alpha^{(k_i)} \to \alpha^{(k_i+1)} \to \cdots \to \alpha^{(k_{i+1})}$$

which satisfies the following conditions $(i = 0, 1, 2, \cdots, m-1)$ respectively:

(i) for every $k_i \leq k < k_{i+1}$, at least one of $(k|k+1)$-newborn letters is a $(k+1|k+2)$-operating letter.

(ii) for $k = k_{i+1}$, the condition is not satisfied.

(6.2) It is easy to see that for every P there corresponds a unique PSCOP:

$$\alpha^{(k_0)} \to \alpha^{(k_1)} \to \alpha^{(k_2)} \to \cdots \to \alpha^{(k_m)} \quad (0 = k_0 < k_1 < \cdots < k_m = n)$$

Here we define the grade of each letter occurrence in $\alpha^{(k_i)}$ $(i = 0, 1, 2, \cdots, m)$ by induction as follows:[1]

[1]Note that grade of letter occurence in α^k $(k \neq k_i)$ is not defined in this paper.

(i) Each letter occurrence a in $\alpha^{(0)} = \alpha^{(k_0)}$ is defined to be grade 0 (written as $G(a) = 0$).

(ii) For any $0 \leq i < m$, each $(k_i|k_{i+1})$-newborn letter occurrence grade in $\alpha^{(k_{i+1})}$ increases up to $G(b)+1$, where b is the biggest grade of letter occurrence among $(k_i|k_{i+1})$-operating letter occurrences of $\alpha^{(k_i)}$; and the $(k_i|k_{i+1})$-reserved letter occurrences of $\alpha^{(k_{i+1})}$ keep their grade unchanged as their corresponding letter occurrences of $\alpha^{(k_i)}$ respectively.

On the above basis, we define the grade of step computation:

$$\alpha^{(k_i)} \longrightarrow \alpha^{(k_{i+1})}$$

of the PSCOP as the biggest grade of all $(k_i|k_{i+1})$-operating letter occurrences in $\alpha^{(k_i)}$.

Hence the classification $I_l (l = 0, 1, 2, \cdots, N)$ of

$$\{i | i = 0, 1, 2, \cdots, m-1\}$$

will be obtained by the following condition: the grade of $\alpha^{(k_i)} \longrightarrow \alpha^{(k_{i+1})}$ is l iff $i \in I_l (l = 0, 1, 2, \cdots, N)$ respectively.

4 The program of parallel computation simulating sequential computation

1^0. The primary condition of the given US computing machine

Let a US computing machine M $= \{A, W, Q, Q_h, \tilde{F}, P\}$ be given with primary condition:

(i) primary word

$$\alpha^{(0)} = a_1^{(0)} a_2^{(0)} \cdots a_{n^{(0)}}^{(0)} \in W$$

and

$$\lambda^{(0)} = \{\dot{i}_1^{(0)}, \dot{i}_2^{(0)}, \cdots, \dot{i}_{k^{(0)}}^{(0)} \quad (1 \leq k^{(0)} \leq n^{(0)})$$

(ii) primary state $q^{(0)} \in Q$.

2^0. The serial computation P of the word

Starting from the corresponding primary word

$$\alpha^{(0)}_{\lambda^{(0)}} = a_1^{(0)} \cdots \overset{\downarrow}{a}{}^{(0)}_{i_1} \cdots \overset{\downarrow}{a}{}^{())}_{i_2} \cdots \overset{\downarrow}{i}{}^{(0)}_{k^{(0)}} \cdots a^{(0)}_{n^{(0)}}$$

of length $n^{(0)}$ and state $q^{(0)}$, we proceed with the serial computation of the word:

$$\alpha^{(0)}_{\lambda^{(0)}} \xrightarrow{\tilde{G}_{q^{(0)}}} \alpha^{(1)}_{\lambda^{(1)},\mu^{(1)}} \xrightarrow{\tilde{G}_{q^{(1)}}} \alpha^{(2)}_{\lambda^{(2)},\mu^{(2)}} \rightarrow \cdots \xrightarrow{\tilde{G}_{q^{(n-1)}}} \alpha^{(n)}_{\lambda^{(n)},\mu^{(n)}}$$

by the given rule $\tilde{F}$:

$$(\alpha_{\lambda,\mu}, q) \stackrel{\tilde{F}}{\longrightarrow} (\alpha'_{\lambda',\mu'}, q'),$$

and so the following informations will be calculated successively: for $k = 0, 1, 2, \cdots, n-1$, $i = 0, 1, 2, \cdots, m-1$,

(1) $\tilde{G}_{q^{(k)}} = (k|k+1)$;

(2) $(k|k+1)$-operating (nonoperating) letter occurrences, $(k|k+1)$-newborn (reserved) letter occurrences;

(3) k_i;

(4) $(k_i|k_{i+1}) = \tilde{G}_{q^{(k_i)}} \cdot \tilde{G}_{q^{(k_i+1)}} \cdots \tilde{G}_{q^{(k_{i+1})}}$;

(5) $(k_i|k_{i+1})$-operating (nonoperating) letter occurences, $(k_i|k_{i+1})$-newborn (reserved) letter occurrences;

(6) the grade of letter occurrences in $\alpha^{(k_i)}$;

(7) the grade of $\alpha^{(k_i)} \to \alpha^{(k_{i+1})}$;

(8) $I_l (l = 0, 1, 2, \cdots, N)$.

3^0. <u>The first part of parallel computation of word</u>

We calculate in parallel for

$$\alpha^{(k_i)} \stackrel{(k_i|k_{i+1})}{\longrightarrow} \quad \alpha^{(k_{i+1})} \quad (i \in I_0)$$

with the corresponding word as follows:

(i) $s(\alpha^{(k_i)}) \stackrel{(k_i|k_{i+1})}{\longrightarrow} \quad T(\alpha^{(k_{i+1})})$ $(i \in I_0)$;

(ii) each letter occurrence of $\overline{T}(\alpha^{(k_{i+1})})$ with grade 0 is transferred from the corresponding letter occurrence in $\alpha^{(k_i)} (i \in I)$ respectively;

(iii)each letter occurrence of $\overline{T}(\alpha^{(k_{i+1})}) (i \in I_0)$ with grade > 0 is kept unchanged.

It is not difficult to see that

(1) all letter occurrences of $\alpha^{(k_i)} (i = 0, 1, 2, \cdots, m-1)$ with grade 0 have been prepared in the given $\alpha^{(0)}$;

(2) all letter occurrences of $\alpha^{(k_i)} (i = 0, 1, 2, \cdots, m-1)$ with grade 1 are included in $T(\alpha^{(k_{i+1})})$ & all letter occurrences of $\overline{T}(\alpha^{(k_{i+1})})$ with grade > 0 must be grade $> 1 (i \in I_0)$.

4^0. <u>The second part of parallel computation of the word</u>.

We calculate in parallel for

$$\alpha^{(k_i)} \overset{(k_i|k_{i+1})}{\longrightarrow} \alpha^{(k_{i+1})} \quad (i \in I_1)$$

with the corresponding word as follows:

(i) $S(\alpha^{(k_i)}) \overset{(k_i|k_{i+1})}{\longrightarrow} T(\alpha^{(k_{i+1})})(i \in I_1)$;

(ii) each letter occurrences of $\overline{T}(\alpha^{(k_{i+1})})$ with grade ≤ 1 is transferred from the corresponding letter occurrence in $\alpha^{(k_i)})(i \in I_1)$ respectively;

(iii) each letter occurrence of $\overline{T}(\alpha^{(k_{i+1})})(i \in I_1)$ with grade > 1 is kept unchanged.

It is not difficult to see that

(1) all letter occurrences of $\alpha^{(k_i)}(i = 0, 1, 2, \cdots, m-1)$ with grade ≤ 1 have been prepared in the first part of parallel computation;

(2) all letter occurrences of $\alpha^{(k_i)}(i = 0, 1, 2, \cdots, m-1)$ with grade 2 are included in $T(\alpha^{(k_{i+1})})$ and all letter occurrences of $\overline{T}(\alpha^{(k_{i+1})})$ with grade > 1 must be grade > 2 $(i \in I_1)$.

......

5^0. The last part of parallel computation of word.

Go on with the above parallel computation, part by part, till we arrive at the last part,

$$\alpha^{(k_i)} \overset{(k_i|k_{i+1})}{\longrightarrow} \alpha^{(k_{i+1})} \quad (i \in I_N)$$

which consists of a single computation with the corresponding word:

$$\alpha^{(k_{m-1})} \overset{(k_{m-1}|k_m)}{\longrightarrow} \alpha^{(k_m)} = \alpha^{(n)}$$

Note that the grades of letter occurrences in $\alpha^{(k_{m-1})}$ are all $\leq N$ and so all letter occurrences in $\alpha^{(k_{m-1})}$ have to be prepared in the former parts of parallel computations.

References

[1] Hu Guoding, On mathematical model of computing machine. J. Computer Sci. and Tech., Vol.3, No.4, pp.273-288, 1988.

[2] Chen Qi, The analysis of Function of Uniform separable MMCM. Master Thesis, Nankai University, 1988.

[3] Zhao Jie, Simulating Uniform Separable Computation With Petri Net-like Languages, Proc. of the Second Asian Symposium, 1990.

[4] Xu Shuruen, Petri Net-like Languages. Proc. of the Second Asian Symposium, 1990.

Computer Mathematics

Proc. of the Special Program at
Nankai Institute of Mathematics
January 1991 – June 1991

An Algorithm and Its Data Structure from Sequential *US MMCM* to a Parallel Machine

Hu Guoding & Wang Yongge
Nankai Institute of Mathematics
Nankai University, Tianjin, P.R.China

Abstract

In [1, 2], Hu Guoding gave a general definition of computing machine model *MMCM*, and proved that one kind of this machine, called *US MMCM*, can be implemented parallelly. In this paper, we will give an algorithm to carry out this transformation from a sequential *US MMCM* to a parallel *US MMCM*, and describe the software TR.LSP of this transformation on an IBM PC/AT.

1 Computing Machine Model *MMCM*

1.1 Notations

Let B be any set, N be a natural number, $< B >_N = \{\alpha | \alpha \in B^*, |\alpha| \leq N\}$.

If $v = a_1 a_2 a_3 \cdots a_n, a_i \in A$, then ${}^{\downarrow}v = \overset{\downarrow}{a}_1\ a_2 a_3 \cdots a_n$, and ${}^{\downarrow}v = \epsilon$ for $n = 0$; $v^{\downarrow} = a_1 a_2 a_3 \cdots \overset{\downarrow}{a}_n$, and $v^{\downarrow} = \epsilon$ for $n = 0$; $\overset{\downarrow\downarrow}{v} = \overset{\downarrow}{a}_1\ a_2 a_3 \cdots \overset{\downarrow}{a}_n$, and $\overset{\downarrow\downarrow}{v} = \epsilon$ for $n = 0$.

1.2 Computing Machine Model *MMCM*

For the detailed information of *MMCM*, the reader is referred to [1]. Here we will give a concise description of *MMCM*.

Definition: An *MMCM* is a 6-tuple (A, W, Q, Q_h, F, P) where

(1) A is a finite alphabet set;

(2) $W \subset A^*$ is the initial word set;

(3) Q is the finite states set and $Q_h \subset Q$ is the halting states set;

(4) $F : E_1 \times (Q - Q_h) \longrightarrow E_2 \times Q$; where $E_1 \subset < A' \cup V >_N, E_2 \subset < A \cup A' \cup V' >_N$ and $V = \{v_1, v_2, \cdots, v_n\}, V' = \{\overset{\downarrow}{v}, \overset{\downarrow}{v}, \overset{\downarrow\downarrow}{v} \mid v \in V\}, A' = \{\overset{\downarrow}{a} \mid a \in A\}$, the elements of E_1 are in the form of $\bar{\alpha} = v_1 \overset{\downarrow}{a_1} v_2 \overset{\downarrow}{a_2} \cdots v_m \overset{\downarrow}{a_m} v_{m+1}; 2m + 1 \leq N$. If $F(\bar{\alpha}, q) = (\bar{\alpha}_1, q_1)$, then $(\bar{\alpha}_1, q_1)$ is only determined by $a_1, a_2, \cdots, a_m, q$; if v or $\overset{\downarrow}{v}, \overset{\downarrow}{v}, \overset{\downarrow\downarrow}{v}$ ocurr in $\bar{\alpha}_1$, then v must occur in $\bar{\alpha}$.

(5) P is the computation processes set, e.g. the following is a computation process:

$$(\alpha_{\lambda_0}, q_0) \rightarrow (\alpha_{\lambda_1}, q_1) \rightarrow \cdots \rightarrow (\alpha_{\lambda_t}, q_t).$$

Now, we will show how to implement one step of this computation:

$$(\alpha_{\lambda_i}, q_i) \rightarrow (\alpha_{\lambda_{i+1}}, q_{i+1}); \quad \alpha_{\lambda_i} \in (A \cup A')^*$$

Firstly, let $\alpha_{\lambda_i} = a_1 \cdots \overset{\downarrow}{a}_{i_1} \cdots \overset{\downarrow}{a}_{i_m} \cdots a_n$ and $v_1 = a_1 a_2 \cdots a_{i_1 - 1}; v_2 = a_{i_1+1} \cdots a_{i_2-1}; \cdots; v_m = a_{i_{m-1}} \cdots a_{i_m - 1}; v_{m+1} = a_{i_m+1} \cdots a_n$. Thus, we have $\bar{\alpha}_{\lambda_i} = v_1 \overset{\downarrow}{a}_{i_1} v_2 \cdots v_m \overset{\downarrow}{a}_{i_m} v_{m+1}$. Using F, we can have $F(\bar{\alpha}_{\lambda_i}, q_i) = (\bar{\alpha}_{\lambda_i'}, q)$. Replacing the v_i in $\bar{\alpha}_{\lambda_i}$ with the above definitions of v_i, we get $\alpha_{\lambda_{i+1}}$; and $q' = q_{i+1}$.

Lastly, let q_0 be the initial state and $q_t \in Q_h$. If we remove the arrows in α_{λ_0}, it becomes an element of W.

2 Uniformly Separable Machine *US MMCM*

Definition: An $MMCM$ M is a uniformly separable $MMCM$, if its instruction set $F : E_1 \rightarrow E_2$ (or we can write it as $(v_1 \overset{\downarrow}{a_1} v_2 \cdots v_m \overset{\downarrow}{a_m} v_{m+1}, q) \overset{F}{\rightarrow} (\bar{\alpha}, q)$) satisfies the following conditions:

(1) the length of $\bar{\alpha}$ depends only on m, q;

(2) q' depends only on m, q;

(3) the letters of $(V \cup V')$ which occur in $\bar{\alpha}$ and their occurring positions in $\bar{\alpha}$ depend only on m, q;

(4) the occurring-positions of letters of A in $\bar{\alpha}$ depend only on m, q;

(5) which letters of $A \cup A'$ will occur in $\bar{\alpha}$ depends only on $a_1, a_2, \cdots, a_m, m$ and q.

3 The Transformation Algorithm from Sequential *US MMCM* to Parallel *US MMCM*

3.1 The data structure of storing a *US MMCM*

Although an *MMCM* M is a 6-tuple (A, W, Q, Q_h, F, P), we need only to store W, Q_h and F, because A can be a default general finite alphabet set, Q belongs to F, and P can be computed from W, F and Q_h. Without loss of generality, we can assume that there is only one element in Q_h and it is stored in the form of an alphabet string in computer. The computation process P of a *US MMCM* is always the same when the length of input word and the operating letters coordinates are the same, so for an N-length input words with the operating letters coordinates $(\lambda_1 \ \lambda_2 \ \cdots \ \lambda_s)$, we can store it in a variable (((a 1) (a 2) (a 3) $\cdots$ (a N) $(\lambda_1 \ \lambda_2 \cdots \lambda_s)$)). For the convenience of LISP language, we use (a i) to denote the variable a_i. Any one production rule of F can be stored in the form of structure(which is equivalent to a record in Pascal language). That is, a production structure is made up of the fields *premise, poperating, pstate, result, roperating, rnewborn* and *rstate*, here *premise* is the premise, stored in the form of ((a 1) (a 2) (a 3) $\cdots$ (a n)); *poperating* is the operating letters coordinates, stored in the form of $(i_1 \ i_2 \cdots i_s)$; *result* is the result which has the same data structure as *premise*; *roperating* is operating letters coordinates for the next operation, stored in the form of $((i_1 \ s_1) \ (i_2 \ s_2) \ \cdots \ (i_t \ s_t))$, where i_j are the operating letters coordinates and $s_j = 0$, 1, 2 or 3, with $(i_j \ 0)$, $(i_j \ 1)$, $(i_j \ 2)$ and $(i_j \ 3)$ denoting $v, \overset{\downarrow}{v}, \overset{\downarrow}{v}$ *and* $\overset{\downarrow\downarrow}{v}$ respectively; *rnewborn* is newborn words, stored in the form of $(\mu_1 \ \mu_2 \ \cdots \ \mu_t)$; *pstate* and *ratate* are states, stored in the form of alphabet string. After the above structural process, the whole F can then be stored in an array *proarry* which has structure as its components.

3.2 The algorithm to obtain *US MMCM*'s computation process

Generally, production rule is in the form of $v_1 c_1 v_2 c_2 \cdots v_n c_n v_{n+1} \rightarrow v_{i_1} b_1 v_{i_2} b_2 \cdots v_{i_m} b_m v_{i_{m+1}}$, where v_i are string variables and c_i, b_i are letters. Because the initial word is $a_1 a_2 \cdots a_n$, to obtain the computation, we must replace subwords between operating letters with the variables, i.e. if $\alpha_0 = a_1 \cdots \overset{\downarrow}{a}_{i_1} \cdots \overset{\downarrow}{a}_{i_m} \cdots a_n$, then let $v_1 = a_1 a_2 \cdots a_{i_1 - 1}; v_2 = a_{i_1+1} \cdots a_{i_2-1}; \cdots; v_m = a_{i_{m-1}} \cdots a_{i_m - 1}; v_{m+1} = a_{i_m+1} \cdots a_n$, so we have $\alpha_0' = v_1 \overset{\downarrow}{a}_{i_1} v_2 \cdots v_m \overset{\downarrow}{a}_{i_m} v_{m+1}$. After doing this, we use the appropriate production rule in F to carry out the computation, and then replace v_i's with the old value. This computation process is the task of function *adapt* and *huanyan* in the software **TR.LSP**. The complexities of *adapt* and *huanyan* are $O(n)$ and $O(n^2)$ respectively. From the above analysis, each one step of the computation process P can be obtained from the following computations:

(1) *adapt*;

(2) search for the appropriate production rule(which is completed by the function *search*);

(3) *huanyan*.

3.3 Find out the maximal parallelism in P

Step 1: Find the $PSCPP$'s in P(Pure Serial Computation Partial Process), i.e. the $k_1, k_2, \cdots, k_t$ in [2]. To do this, for each step of P, we test whether there are any operating letters of this step which belong to the newborn words of the last step. The result of this step is stored in the array *klevel*, i.e. *klevel*$[i]$ is made up of the steps of P which are in k_i. The complexity of this step is $O(n)$.

Step 2: Find out the maximum parallelism for each k_i in $PSCPP$, i.e. we classify $PSCPP$ into $I_1, I_2, \cdots, I_s$ so that for each I_j, the elements k_i in I_j can be implemented parallelly. To get I_j, firstly, we must have the operating letters for each k_i. (Remark: k_i is a composite map of simple $US\ \ MMCM$ maps. Its operating letters are the union of the submaps' operating letters.) Secondly, use the above result to label each k_i, e.g. if the operating letters of k_i are produced by $k_{j_1}, k_{j_2}, \cdots, k_{j_t}$, then give k_i the label $\max(j_1, j_2, \cdots, j_t)$. Lastly, use the labels to get the classification. The complexity of step 2 is $O(n^3)$.

4 The Auto-transformation Software From Sequential Algorithm to a Parallel Algorithm

We have designed a software **TR.LSP** which can be used on an IBM PC/AT to carry out this transformation. For a sequential $US\ \ MMCM$, the software give the parallel machine which is equivalent to the input sequential machine, i.e., give the computation process P:

$$\alpha_0 \xrightarrow{1*} \alpha_1 \xrightarrow{2*} \alpha_2 \xrightarrow{3*} \cdots \xrightarrow{n*} \alpha_n$$

and classify this process into: $I_1, I_2, \cdots, I_l$ which satisfies:

(1) $\cup_{i=1}^{l} I_i = \{1*, 2*, \cdots, n*\}$;

(2) $I_i \cap I_j = \phi, i \neq j$;

(3) If $I_i = \{n_1*, n_2*, \cdots, n_k*\}$, then, in the process P, the $n_1*, n_2*, \cdots, n_k*$-th step can be implemented parallelly.

In other words, the output parallel machine operates as follows:

Step 1. Carry out each step in I_1 parallelly;

Step 2. Carry out each step in I_2 parallelly;

… …

Step 1. Carry out each step in I_l parallelly; and output the whole result.

The structural scheme of this software can be found at the end of this article.

5 An application of This Software

Example: One application of software TR.LSP: Multiplication of numbers:

(I) The production rules for $US\ \ MMCM$ M of multiplication

(((v 1) (a 1) (a 2) (a 3)) (2 3 4) "q0" ((v 1) (b 1) (b 2) (a 2) (a 3) (b 3) (b 4)) ((1 2) (4 0) (5 0)) (2 3 6 7) "q1")

(((v 1) (a 1) (v 2) (a 2) (a 3) (v 3)) (2 4 5) "q1" ((v 1) (b 1) (b 2) (v 2) (a 2) (a 3) (b 3) (b 4) (v 3)) ((1 2) (5 0) (6 0)) (2 3 7 8) "q1")

(((a 1) (v 1) (a 2) (a 3) (v 2)) (1 3 4) "q1" ((b 1) (b 2) (v 1) (a 3) (b 3) (b 4) (v 2)) ((3 2) (4 0)) (1 2 5 6) "q1")

(((v 1) (a 1) (a 2) (v 2)) (2 3) "q1" ((v 1) (a 1) (a 2) (v 2)) ((1 2) (3 0)) nil "q1")

(((v 1) (a 1) (v 2) (a 2) (v 3)) (2 4) "q1" ((v 1) (a 1) (v 2) (a 2) (v 3)) ((1 2) (2 0) (4 0)) nil "q2")

(((v 1) (a 1) (v 2) (a 2) (v 3)) (2 4) "q2" ((v 1) (a 1) (v 2) (a 2) (v 3)) ((1 2) (2 0) (4 0)) nil "q2")

(((v 1) (a 1) (a 2) (v 2) (a 3) (v 3)) (2 3 5) "q2" ((v 1) (b 1) (b 2) (v 2) (a 3) (v 3)) ((1 2) (5 0)) (2 3) "q2")

(((a 1) (v 1) (a 2) (v 2)) (1 3) "q2" ((a 1) (v 1) (a 2) (v 2)) ((2 2) (3 0)) nil "q3")

(((a 1) (a 2) (v 1) (a 3) (v 2)) (1 2 4) "q2" ((b 1) (b 2) (v 1) (a 3) (v 2)) ((3 2) (4 0)) (1 2) "q3")

(((v 1) (a 1) (a 2) (v 2)) (2 3) "q3" ((v 1) (a 1) (a 2) (v 2)) ((1 2) (3 0)) nil "q3")

(((v 1) (a 1) (v 2) (a 2) (v 3)) (2 4) "q3" ((v 1) (a 1) (v 2) (a 2) (v 3)) ((1 2) (4 0)) nil "q4")

(((v 1) (a 1) (v 2) (a 2) (v 3)) (2 4) "q4" ((v 1) (a 1) (v 2) (a 2) (v 3)) ((1 2) (4 0)) nil "q5")

(((v 1) (a 1) (v 2) (a 2) (v 3)) (2 4) "q5" ((v 1) (a 1) (v 2) (a 2) (v 3)) ((1 2) (2 0) (4 0)) nil "q5")

(((v 1) (a 1) (a 2) (v 2) (a 3) (v 3)) (2 3 5) "q5" ((v 1) (b 1) (v 2) (a 3) (v 3)) ((1 2) (4 0)) (2) "q5")

(((a 1) (v 1) (a 2) (v 2)) (1 3) "q5" ((b 1) (v 1) (a 2) (v 2)) ((3 0) (4 2)) (1) "q7")

(((a 1) (a 2) (v 1) (a 3) (v 2)) (1 2 4) "q5" ((b 11) (v 1) (a 3) (v 2)) ((3 0) (4 2))

(1) "q7")

(((v 1) (a 1) (v 2) (a 2)) (2 4) "q7" ((v 1) (a 1) (v 2) (a 2)) ((2 0) (3 2)) nil "q7")

(((v 1) (a 1) (v 2) (a 2) (v 3)) (2 4) "q7" ((v 1) (a 1) (v 2) (a 2) (v 3)) ((2 0) (3 2) (4 0)) nil "q8")

(((v 1) (a 1) (v 2) (a 2) (v 3)) (2 4) "q8" ((v 1) (a 1) (v 2) (a 2) (v 3)) ((2 0) (3 2) (4 0)) nil "q8")

(((v 1) (a 1) (v 2) (a 2) (a 3) (v 3)) (2 4 5) "q8" ((v 1) (a 1) (v 2) (b 7) (b 8) (v 3)) ((2 0) (3 2)) (4 5) "q8")

(((v 1) (a 1) (a 2) (v 2)) (2 3) "q8" ((v 1) (a 1) (a 2) (v 2)) ((2 0) (4 2)) nil "q9")
Ω(((v 1) (a 1) (a 2) (a 3) (v 2)) (2 3 4) "q8" ((v 1) (a 1) (b 1) (b 2) (v 2)) ((2 0) (5 2)) (3 4) "q9")

(((v 1) (a 1) (v 2) (a 2)) (2 4) "q9" ((v 1) (a 1) (v 2) (a 2)) ((2 0) (3 2)) nil "q9")

(((v 1) (a 1) (v 2) (a 2) (v 3)) (2 4) "q9" ((v 1) (a 1) (v 2) (a 2) (v 3)) ((2 0) (3 2)) nil "q10")

(((v 1) (a 1) (v 2) (a 2) (v 3)) (2 4) "q10" ((v 1) (a 1) (v 2) (a 2) (v 3)) ((2 0) (3 2)) nil "q11")

(((v 1) (a 1) (v 2) (a 2) (v 3)) (2 4) "q11" ((v 1) (a 1) (v 2) (a 2) (v 3)) ((2 0) (3 2) (4 0)) nil "q11")

(((v 1) (a 1) (v 2) (a 3) (a 4) (v 3)) (2 4 5) "q11" ((v 1) (a 1) (v 2) (b 9) (v 3)) ((2 0) (3 2)) (5) "q11")

(((v 1) (a 1) (a 2) (v 2)) (2 3) "q11" ((v 1) (a 1) (b 1) (v 2)) ((1 2) (2 0) (4 2)) (3) "q12")

(((v 1) (a 1) (a 2) (a 3) (v 2)) (2 3 4) "q11" ((v 1) (a 1) (b 11) (v 2)) ((1 2) (2 0) (4 2)) (3) "q12")

(((v 1) (a 1) (a 2) (v 2) (a 3)) (2 3 5) "q12" ((v 1) (a 1) (a 2) (v 2) (a 3)) ((1 2) (3 0) (5 0)) nil "q13")

(((v 1) (a 1) (v 2) (a 2) (v 3) (a 3)) (2 4 6) "q13" ((v 1) (b 1) (b 2) (v 2) (a 2) (v 3)) ((1 2) (5 0) (6 2)) (2 3) "q13")

(((a 1) (v 1) (a 2) (v 2) (a 3)) (1 3 5) "q13" ((b 1) (b 2) (v 1) (a 2) (v 2)) ((4 0) (5 2)) (1 2) "q13")

(((v 1) (a 1) (v 2) (a 2)) (2 4) "q13" ((b 1) (b 2) (v 1) (a 1) (v 2)) ((4 0) (5 2)) (1 2) "q13")

(((v 1) (a 1) (a 2)) (2 3) "q13" ((b 1) (b 2) (v 1) (a 1)) ((3 2) (4 0)) (1 2) "q14")

(((v 1) (a 1) (a 2)) (2 3) "q14" ((v 1) (a 1)) ((1 2)) nil "q14") (((v 1) (a 1) (v 2)) (2) "q14" ((v 1) (a 1) (v 2)) ((1 2)) nil "q15")

(((v 1) (a 1) (v 2)) (2) "q15" ((v 1) (a 1) (v 2)) ((1 2) (2 0)) nil "q15") (((v 1) (a 1) (a 2) (v 2)) (2 3) "q15" ((v 1) (b 1) (v 2)) ((1 2)) (2) "q15")

(((a 1) (a 2) (v 1)) (1 2) "q15" ((b 1) (v 1)) nil (1) "qe") (((a 1) (v 1)) (1) "q15" ((b 1) (v 1)) nil (1) "qe")

nil

(II) The computation process for the above $MMCM$ when input length "2" × "10" is

$$\alpha_0 \xrightarrow{1*} \alpha_1 \xrightarrow{2*} \cdots \xrightarrow{128*} \alpha_{128}$$

The exact description of α_i is omitted here.

(III) The parallel machine for multiplication of "10" × "2" is

I_1 = (1 2 3 4 5 6 7 8 9 10)

I_2 =(11 12 13 14 15 16 17 18 19 20 21 22 23 24 25 26 27 28 29 30 31 50 51 52 53 54 55 56 57 58 59 60 61 62 63 64 65 66 67 68 69 70 71 90 91 104)

I_3 =(32 33 34 35 36 37 38 39 40 41 42 43 44 45 46 47 48 49 72 73 74 75 76 77 78 79 80 81 82 83 84 85 86 87 88 89 92 93 94 102 103)

I_4 =(95 96 97 98 99 100 101 105 106 107 108 109 110 125 126 127)

I_5=(111 112 113 114 115 116 117 118 119 120 121 122 123 124 128)

The above result shows that for the multiplication of form "10" × "2", the sequential machine needs 128 steps, but the parallel machine needs only 5 steps: I_1, I_2, I_3, I_4, I_5.

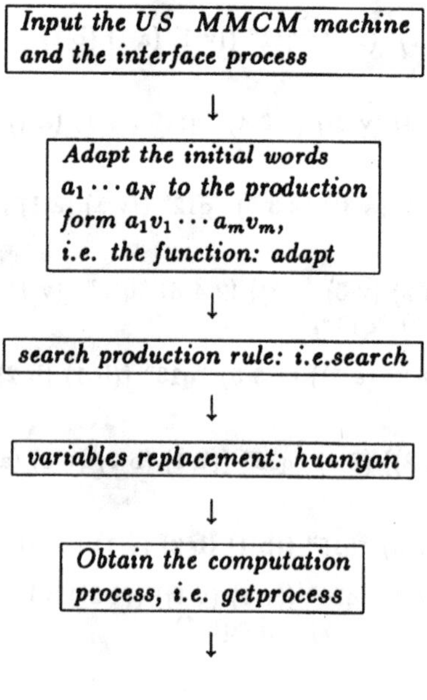
Input the US MMCM machine and the interface process
Adapt the initial words $a_1 \cdots a_N$ to the production form $a_1 v_1 \cdots a_m v_m$, i.e. the function: adapt
search production rule: i.e.search
variables replacement: huanyan
Obtain the computation process, i.e. getprocess

Label the PSCPP
i.e. getlevel

↓

Classification: getIs

↓

The mainbody of software

↓

Show the results

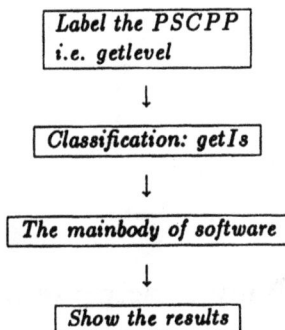

Fig.1 The structural scheme of TR.LSP

References

[1] Hu Guo-ding. On Mathematical Model of Computing Machine, J. of Comp. 1988.

[2] Hu Guo-ding. Parallel Machine Simulating Serial Computing Machine, *in this volume.*

[3] Gold Hill Computer Company, Golden Common Lisp Reference.

[4] L.Silklossy, Let's Talk Lisp, 1976.

Computer Mathematics
Proc. of the Special Program at
Nankai Institute of Mathematics
January 1991 - June 1991

Sketch of a New Discipline of Modelling

E. Engeler
ETH Zürich

0. Introduction

The most successful discipline of modolling is clearly that of differential equations: it provides the cognitive tools to recognize the structure of a system of (physical) processes; it provides a mathematical environment suited to treat local as well as global aspects of the system; and it comes with a highly developed set of numerical techniques for the actual evaluation of the models for determining parameters, discriminating between modelling choices and making quantitative predictions. To provide these possibilities to the scientist makes the field of differential equations a discipline in the present sense. Any other proposal for a discipline of modelling will have to submit to a comparison with that well-tested one.

While we cannot, obviously, compete with differential equations in listing striking successes for our proposal, we must show how to address the three main aspects mentioned above. The cognitive tool is provided by interaction graphs, (Section 1); the mathematical environment is the theory of combinatory algebras, (Section 2); and the solution technology is described by a directed search procedure, (Section 3).

We shall now give an outline of our approach. To fix ideas, let us consider the case of differential equations in a new light: an equation such as $y = D(y) + f(x)$ relates variables of differential type, e.g. real variables, variables f, g denoting functions on reals, and differential operator D. Each such variable denotes a mathematical object. Conventionally this denotation is a specific real number, a given function, or a functional. There is room to relax this convention somewhat: instead of denoting exact values, a variable may denote some knowledge about that value, knowledge expressed in an appropriate language. For example "$2 < @ < 3$", is a formula which we understand to express the fact that the value of the variable under discussion is between two and three. In the case of functions "$@ = \lambda t \cdot sint$" would state precisely that the function variable denotes the sine function, while "$periodic(@)$", "$-1 < @ < 1$" would be formulas expressing (partial) knowledge about this function.

If a function f is applied to a variable x then it transforms knowledge about the arguments into knowledge about the result; e.g. sine transforms "$0, @, \pi/4$" into "$0 < @ < \sqrt{2}/2$". This fact is also knowledge (about the sine function), and we shall represent this kind of knowledge by formal expressions using arrow:

$$\{0 < @ < \pi/4\} \longrightarrow 0 < @ < \sqrt{2}/2$$

would be a formula added to our set of knowledge about sine, together with "$-1 < @ < 1$" etc.

The application of (knowledge about) a function to (knowledge about) a variable thus becomes an operation on sets of formulas. The basic thesis on the computability of such set operations is that they are completely determined by their action on finite sets,or, as it is conveniently stated, that they are continuous. Thus, the function f gives rise to a set operation F with

$$F(M) = \cup\{F(\alpha) : \alpha \subseteq M, \alpha \text{ finite}\}.$$

We have already proposed to represent knowledge about f, hence about F, as a set of formulas of the form $\alpha \longrightarrow a$, where α is a finite set, and a is a member, of a set of formulas expressing basic properties of objects "at the ground level". If we consider F to be this set then the application of F to a set M is an operation on sets which we write multiplicatively and define as

$$F \cdot M := \{a : \exists \alpha \subseteq M.\ \ \alpha \longrightarrow a \in F\}.$$

Altogether, F is therefore a set of formulas $\{b_1, \ldots, b_n\} \longrightarrow a$ such that whenever the arguments of the function f have properties $b_1, \ldots, b_n$ then the value of f will have property a.

What was said about functions can also be said about operations on functions, e.g. differentiation. Thus, for example, the differentiation operation D corresponds to a set of formulas which contain

$$\{periodic(@), -1 < @ < 1\} \longrightarrow periodic(@)$$
$$\{linear(@)\} \longrightarrow @ = 0, \text{ etc.}$$

For the representation of operations with more than one argument, we can make use of Schönfinkels device. Let, for example, g be a binary operation about which we know that if the first argument has properties $b_1, \ldots, b_m$ and the second one has $c_1, \ldots, c_n$ then the value has property a. Then g will be represented by a set G of expressions containing the corresponding element

$$\{b_1, \ldots, b_m\} \longrightarrow (\{c_1, \ldots, c_n\} \longrightarrow \alpha).$$

The application of G is then simply $(G{\cdot}M){\cdot}N$, where $\{b_1, ..., b_m\} \subseteq M$, $\{c_1, ..., c_n\} \subseteq N$.

There are simple examples, where the parentheses in expressions with arrows are differently placed, e.g. $\{\beta \longrightarrow b, v \longrightarrow c\} \longrightarrow a$ instead of $\alpha \longrightarrow (\beta \longrightarrow a)$ as above: consider the following property of the differentiation operation

$$\{differentiable(@), \{@ < 0\} \longrightarrow @ < 0, \{0 < @ < 1\} \longrightarrow @ > 0,$$
$$\{1 < @\} \longrightarrow @ < 0\} \longrightarrow has - zero(@),$$

which expresses the fact that the derivative has a zero if the function has at least two.

By these considerations we are lead to a natural formalism for the representation of knowledge about the objects of our domain. Starting with a vocabulary A of basic facts we build a cumulative hierarchy of sets of expressions $G_n(A)$ by

$$G_0(A) = A,$$

$$G_{n+1}(A) = G_n(A) \cup \{\alpha \longrightarrow a : \alpha \subset G_n(A), \alpha \text{ finite}, a \in G_n(A)\}.$$

The resulting language for knowledge representation is the union of these sets

$$G(A) = \cup_n G_n(A);$$

its subsets are the (representation of) the objects of our discussion. The application of one object M to another N remains of course the same as above

$$M \cdot N := \{a : \exists \alpha \subset N.\alpha \longrightarrow a \in M\}.$$

In this way the set D_A of all subsets of $G(A)$ is furnished with a binary operation which makes it into an algebraic structure, a **combinatory algebra** as it were. Eventually, as we shall see, the proposed discipline will consist of describing the setting of a problem by a set of equations in D_A and then solve them.

1. Interaction Graphs and their Equations

The case of differential equations served us well as an introductory example for the kind of modelling that we intend to develop: we saw how objects of quite different types can be represented as sets of knowledge about them and how the basic interrelation between them, application, can be realized as an operation on sets (of formulas). We shall now abstract from that particular context, retaining only the construction of $G_n(A)$ and D_A from a set A of basic formulas and the operation of application $M.N$ for arbitrary subsets of $G(A)$, i.e. element of D_A. We find it natural to call such sets "processes": they include descriptions of "behavior", "laws determining behavior", etc., always as sets of formulas in an appropriate language $G(A)$. A **process**, then, is simply an element of D_A.

Different types of processes obtain their difference not from their substance, but rather from their function in a network of processes. To visualize this concept, let us now introduce pictograms for processes. Each process is represented by a labeled circle. It is convenient to think of this circle to contain knowledge about that process. The network of interactions between processes is represented by arrows. These are used with the following conventions:

- an arrow terminating at process indicates that this process is being influenced by information carried along the arrows;
- an arrow originating at a process carries away information from that process;
- an arrow, or a bundle of arrows, passing through a process indicates, that this process takes account of the information that it receives and sends away corresponding "processe" information.

Thus, in fig. 1 the process p_0 acts on information received from p_1 and p_2 and influence p_3.

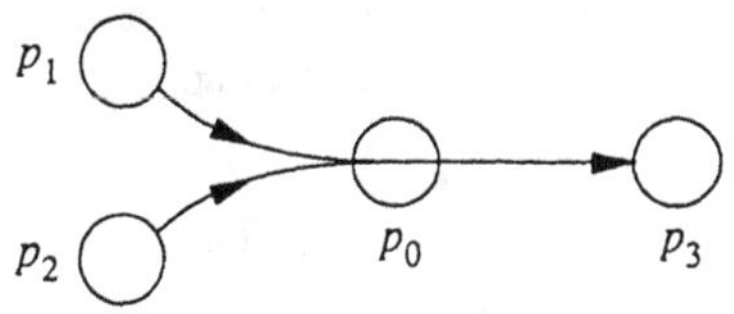

Fig.1 A simple system

Interaction graphs use diagrams as the one above as building blocks. Fig.2 gives an examples of two interacting processes x and y whose mutual dependence is governed by two additional processes f and g.

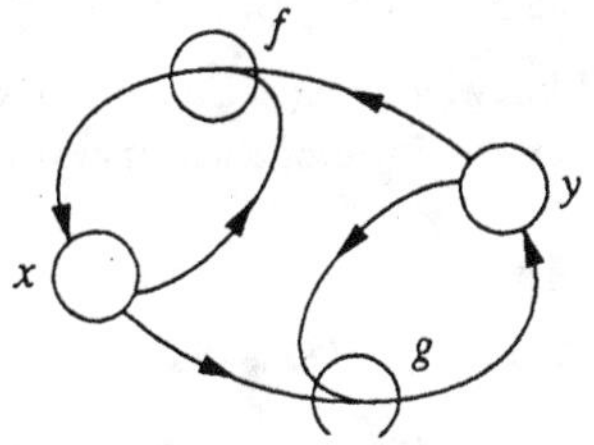

Fig.2 Coupled processes

The process g which determines the dependence of y from x and itself may of course itself be dependent on, say, an additional process z and knowledge about the dependency law embodied in the process f. The augmented intersection graph is depicted in fig.3.

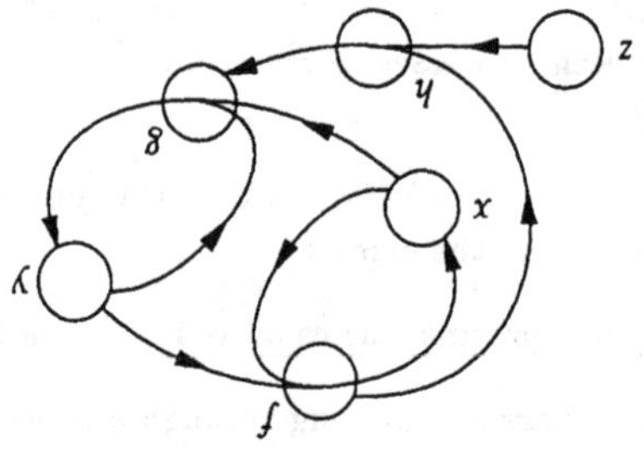

Fig. 3 Hierarchically coupled processes

In this fashion we can represent complex hierarchical and recursive mutual dependencies between processes in a transparent graphical manner. Interaction graphs thus serve at least as a visual help to present and discuss more or less intricate causal dependencies; they are an aid in conceptual modelling.

Let P be a non-empty set of processes and let Q be a set of finite sequence of elements of P, each sequence having length ≥ 3 Then the pair

$$I = < P, Q >$$

is called an (abstract) **interaction graph.** Concrete interaction graphs are obtained by identifying elements p of P as processes in the sense of Section 0, i.e. as subsets of $G(A)$ for some adequate language A. The set Q simply describes the nature of the "arrows": $p_0 p_1 ... p_n p_{n+1} \in Q$ means that p_{n+1} is the terminal node of a bundle of arrows coming from $p_1, ..., p_n$ and all passing through p_0.

Thus, p_0 would contain formulas of the form

$$\alpha_1 \longrightarrow (\alpha_2 \longrightarrow ... \longrightarrow (\alpha_n \longrightarrow \alpha)...)$$

to describe the handling by p_0 of knowledge $\alpha_1, ..., \alpha_n$ obtained from $p_1, ..., p_n$ respectively. This handling has been given the form of a set operation on subsets of $G(A)$, as follows

$$p \cdot q = \{a : \exists \alpha \subset q, \alpha \longrightarrow a \in p\}.$$

With this notion, the element $p_0 p_1 ... p_n p_{n+1}$ of Q is in fact transformed into an equation

$$p_{n+1} = (...(p_0 \cdot p_1) p_2 ... p_n).$$

The question, which laws of interaction (determining p_{n+1} from $p_1, ..., p_n$) can be represented by processes such as p_0 has a well-known answer: it is the case iff the dependenc

$$p_{n+1} = F(p_1, ..., p_n)$$

is continuous in the following sense

$$F(p_1,...,p_n) = \cup_{\alpha_i \subseteq p_i} F(\alpha_1,...,\alpha_n), \quad \alpha_i \text{finite}.$$

Then, in fact we can represent F by p_0, given by

$$p_0 = \{(\alpha_1 \longrightarrow (\alpha_2 \longrightarrow ...(\alpha_n \longrightarrow a)...) : a \in F(\alpha_1,...,\alpha_n), \alpha_i \subseteq p_i, \alpha_i \text{ finite}\}.$$

The above translation of an element $p_0p_1...p_np_{n+1}$ of the interaction graph into an equation suggests a relation between interaction graphs as a whole and a set of equations. To describe this relation effectively, let P be a set of processes and let t be any term built up from atomic symbols p_i (for process in P) by multiplication(and parentheses). We start by defining an interaction graph $\mathbf{I}(t) =< P(t), Q(t) >$ recursively on the structure of t.

If t is atomic, then $\mathbf{I}(t) =< \{t\}, \emptyset >$.
If t is of the form $p \cdot q$ then $\mathbf{I}(t) =< P(t), Q(t) >$, where
$P(t) = P(p) \cup P(q) \cup \{t\}$ and
$Q(t) = Q(p) \cup Q(q) \cup \{pqt\}$.

Thus, new process symbols are introduced for each result of multiplying two terms. In line with our convention of multivaritite dependencies, we now modify the recursion by considering maximal left-associative subterms $p = p_1...p_n$, which contribute only p to $P(t)$ and $p_1...p_n \cdot p$ to $Q(t)$.

Consider now equations Γ, say $t_i{}' = t_i{}'', i = 1, 2, ..., m$, which we assume normalized, i.e. never both $t_i{}'$ and $t_i{}''$ atomic, and if one is atomic then let it be $t_i{}'$. For each equation let t_i be a new atomic process symbol in case both $t_i{}'$ and $t_i{}''$ are composite, and let t_i be $t_i{}'$ in the other case. Using these conventions, the interaction graph associated to Γ is defined by

$$\mathbf{I}(\Gamma) =< P(\Gamma), Q(\Gamma) >,$$

with

$$P(\Gamma) = \{t_i : i = 1, ..., m\} \cup \cup_{i=1}^{m} P(t_i{}') \cup P(t_i{}'');$$

$$Q(\Gamma) = \cup_{i=1}^{m} Q(t_i{}') \,|_{t_i'}^{t_i} \cup \cup_{i=1}^{m} Q(t_i{}'') \,|_{t_i''}^{t_i},$$

where $Q(t_i{}') \,|_{t_i'}^{t_i}$ is the result of substituting t_i for all occurrences of $t_i{}'$ in $Q(t_i')$. In this fashion, every finite set Γ of equations is presented by an interaction graph. The following lemma states the converse and answers the question of what is the relation between Γ and Γ' if $\mathbf{I}(\Gamma) \cong \mathbf{I}(\Gamma')$.

Every finite interaction graph I is isomorphic to an interaction scheme $I(\Gamma)$, where Γ is determined uniquely up to renaming of terms.

Namely, let $I =< P, Q >$ and proceed inductively on the size of Q. If $| Q |= 1$, say $Q = \{s_0s_1...s_ns_{n+1}\}$, let $\Gamma = \{s_{n+1} = s_0 \cdot s_1...s_n\}$. For the induction step, let $Q_{m+1} = Q_m \cup \{s_0...s_ns_{n+1}\}, | Q_m |= m, | Q_{m+1} |= m+1$. Let P_m, P_{m+1} be the process symbols occurring in Q_m, Q_{m+1} respectively, and let $I_m =< P_m, Q_m >$, $I_{m+1} =< P_{m+1}, Q_{m+1} >$. By induction hypothesis there exists Γ_m such that $I_m \cong$ $\mathbf{I}(\Gamma)$. Now let

$$\Gamma_{m+1} = \Gamma_m \cup \{s_0 \cdot s_1...s_n = s_{n+1}\} \cup \{s_0 \cdot s_1...s_n = a_0 \cdot a_1...a_k : a_0...a_ks_{n+1} \in Q_m\}.$$

Then $I(\Gamma_{m+1} \cong< P_{m+1}, Q_m \cup \{s_0...s_{n+1}\} \cup \{a_0...a_ks_{n+1} \in Q_m\} >=< P_{m+1}, Q_{m+1} >$. Finally, assume $I(\Gamma) \cong I(\Gamma')$. Then, with $I(\Gamma) =< P(\Gamma), Q(\Gamma) >$ and $I(\Gamma') =< P(\Gamma'), Q(\Gamma') >$ we have $| P(\Gamma) |=| P(\Gamma') |$, and may hence assume $P(\Gamma) = P(\Gamma')$. Now the equations are read off uniquely from $Q(\Gamma)$ and $Q(\Gamma')$ and are only determined by the isomorphism type. Hence the result.

2. The Combinatory Algebra of Processes

The introduction of a binary operation $p \cdot q$ for subsets p, q of $G(A)$ furnishes the set D_A of subsets of $G(A)$ with an algebraic structure $\mathbf{D}_A =< D_A, \cdot >$. We will often suppress the dot between p and q. It is well-known that this is a combinatory algebra.[1] As such it possesses an element Y which acts as a fixed point operator: for all f we have

$$f(Yf) = Yf;$$

indeed Y can be chosen the least fixed point operator, i.e. an operator which produces

$$Yf = \cup_{n=0}^{\infty} f^{(n)} \cdot \phi, \text{ where}$$

$$f^{(0)} = f, f^{(n+1)} = f \circ f^{(n)}.$$

The least fixpoint that extends a given approximation p_0 to a solution p of $fp = p$ is similarly obtained, namely by

$$p = \cup_{n=1}^{\infty} f^{(n)} \cdot p_0.$$

For simultaneous fixpoint equations, e.g.

$$p = fpq, \quad q = gpq$$

a similar construction can be employed, namely

$$p = \cup p_i, q = \cup q_i, \text{ where } p_{i+1} = fp_iq_i, q_{i+1} = gp_iq_i.$$

Thus, the case of (simultaneous) fixpoint equations is easily understood mathematically and for the case that $f \subseteq G_1(A)$ we have in the backtrack and unification/resolution algorithms of pure logic programs a prototype for algorithmic realization. There is

[1]E. Engeler: "Cumulative Logic Programs and Modelling". In: Logic Colloquium'86, (Drake and Truss, eds.), North-Holland 1988, pp.83-93.

therefore a premium on the formulation of modelling programs which can be given the form of (simultaneous)fixpoint equations. This may, however, not be possible in many cases, and we have to search for a more general approach to this kind of programming by equations. For this purpose it is helpful that the combinatory algebra D_A satisfies the following normal-form lemma[2] : let $a_1, ..., a_k$ be given elements of D_A, i.e. subsets of $G(A)$, and let

$$f_i(a_1, ..., a_k, x_1, ..., x_m) = g_i(a_1, ..., a_k, x_1, ..., x_m), i = 1, ..., m,$$

be a set of equations in D_A for the unknowns $x_1, ..., x_m$. Then there exist elements a and b of D_A such that solving the original set of equations for $x_1, ... x_m$ is equivalent to solving the equations $a \cdot x = b \cdot x$ for x. Indeed, if the sets $a_1, ..., a_k$ are enumerated by some program, then a and b can be effectively constructed and the passage from a solution x to solutions $x_1, ..., x_m$(and conversely) is also effective.

This effectiveness of course does not imply that there is a general solution algorithm for $a \cdot x = b \cdot x$. This fact follows at once from the observation that D_A, being a combinatory algebra, is subject to the unsolvability phenomenon that each universal computation device exhibits. Thus, it is easy to construct recursively enumerable a and b such that no recursively enumerable x solves the equation.

3. Consistency and the Search for Solutions

The result of the detour through algebra taken in the previous section is simple, that we may, in principle, retrict our attention to solving equations of the form

$$a \cdot x = b \cdot x$$

for given a and b.

The identification of process with subsets of $G(A)$ starts with the choice of some space A of basic attributes for processes; this space may be visualized e.g. as set of atomic formulas of some first-order language. In that case, we have the familiar notion of consistency(of subsets $B \subseteq A$), with the property that the empty set is consistent, and that B is consistent iff all its finite subsets are consistent. Using these properties, it is easy to extend the notion of consistency to all the subsets of $G(A)$, namely by recursion on the level of elements: assume that consistency is defined for all finite sets of level $\leq n$ and consider $B = \{a_1, ..., a_n, \alpha_1 \longrightarrow b_1, ..., \alpha_m \longrightarrow b_m\}$ where a_i, b_i, α_i are of level $\leq n$. Then B_0 is said to be consistent if $\{a_1, ..., a_n\}$ is consistent and if, whenever $\alpha_{j_1} \cup ... \cup \alpha_{j_k}$ is consistent, then also $\{b_{j_1}, ..., b_{j_k}\}$ is consistent. In line with the consistency notion at the basic level, we say that $B \subseteq G(A)$ is consistent, if all its finite subsets are.

[2]E. Engeler: "Representation of Varienties in Combinatory Algebras", Algebra Universalis 25 (1988), pp.85-95.

In conception with model construction it occurs occasionally that consistency is not only built up from the lowest level. For example, we want a process p described by the sine function:

$$\{@ = \lambda t \cdot sint, \{@ = 0\} \longrightarrow @ = 0, \{@ = \pi/2\} \longrightarrow @ = 1\}$$

would be considered consistent, but

$$\{@ = \lambda t \cdot sint, \phi \longrightarrow @ > 1\}$$

would certainly be inconsistent. Thus, our definition of consistency will have to be adapted to this situation, preferably by giving the axiomatic requirements for a notion of consistency.

A **Consistency basis** C_0 is a set of finite subsets of $G(A)$ such that $\phi \in C_0$ and such that, if

$$\{(\alpha_i \longrightarrow a_i) : i = 1, ..., n\} \in C_0 \text{ and } \alpha_{j_1} \cup ... \cup \alpha_{j_k} \in C_0 \text{ then } \{\alpha_{j_1}, ..., \alpha_{j_k}\} \in C_0.$$

The **consistency closure** C of C_0 is the smallest set of subsets of $G(A)$ such that $C_0 \subseteq C$ and

(i) $N \in C$ for all finite $N \subseteq M$ implies $M \in C$;

(ii) $N \subseteq M \in C$ implies $N \in C$;

(iii) $\{(\alpha_i \longrightarrow a_i) : i = 1, ..., n\} \in C, \alpha_{j_1} \cup ... \cup \alpha_{j_k} \in C$ implies $\{\alpha_{j_1}, ..., \alpha_{j_k}\} \in C$.

such a closure obviously exists and is uniquely determined by C_0.

Observe, that if C_0 is decidable, then it is decidable for any finite subset $\alpha \subseteq G(A)$ whether α is in the consistency closure C of C_0. In order to make the notion of decidability precise, let us consider elements of $G(A)$ represented as lists(in the sense of LISP) over A as the set of atomic lists. We shall adopt the PROLOG notation for lists, taking $[a \mid \alpha]$ to be the list with head a and tail α. Finite sets of lists are represented as linear lists, and for such lists (α, β) is simply the linear list representing $\alpha \cup \beta$. Therefore we realize the elements $\alpha \longrightarrow a$ of $G(A)$ as listed $[a \mid \alpha]$.

In the following, it is convenient to use the language of logic programming to describe computations with subsets of $G(A)$. The following lemma belongs to the folklore of this subject: for every recursively enumerable set $M \subseteq G(A)$ there is a logic program on lists for a predicate $m(L)$ such that M is the set of all answers λ to queries $? : m(L)$.

As a corollary to this lemma, we have the following consequence of the decidability of the consistency basis C_0: there exist logic programs cons(L) and incons(L) which generate the consistent, resp. inconsistent finite subsets of $G(A)$, (in the form of linear lists).

Let us now consider the problem of solving

$$a \cdot x = b \cdot x$$

where a and b are given, as recursively enumerable sets, by logic programs for the predicates $a(L)$ and $b(L)$ respectively. We seek to approximate a solution x by finding a sequence of consistent expansions x_i of some seed x_0. Each x_i aims to satisfy the equation $a_i x_i = b_i x_i$, where a_i, b_i are those parts of a, b which have been generated up to this point.(To obtain a_{i+1} from a_i query $? : a(L)$ again and the result to a_i.)

At each level of approximation x_i query $? : cons([L \mid x_i])$, where x_i denotes the linear list representing the approximation set x_i. Let this query return the list λ. This is added to x_i, to obtain a provisional next step of approximation x_{i+1}. Now it may happen that $a_{i+1}x_{i+1} \cup b_{i+1}x_{i+1}$ is inconsistent. Then such an expansion x_{i+1} can not be part of a solution: because both ax and bx are consistent and equal for solutions x we would have $a_{i+1}x_{i+1} \cup b_{i+1}x_{i+1} \subseteq ax \cup bx = ax$ and therefore, impossibly, an inconsistent subset of a consistent set. Therefore, in our process of approximation, it is of advantage to identify "window", i.e. sets w such that $aw \cup bw$ is inconsistent.

In the logic programming contexts such windows can be obtained as follows: let $im(L_1, L_2)$ represent the relation $ax_1 \cup bx_1 = x_2$, where L_i are the (linear) list representation of the (finite) sets x_i. Then the following logic program generates windows:
$im([], L) : -,$
$im([U \mid V], L) : -im(V, L), a([U \mid L]),$
$im([U \mid V], L) : -im(V, L), b([U \mid L]),$
$w(L) : -im(U, L), incons(U).$

Thus, after having queried $? : cons([L \mid x_i])$ and having obtained λ we now query $? : w([\lambda \mid x_i])$. If the query returns no then x_i is enlarged by λ and we iterate.

Of course, instead of instituting a blind search $? : cons([L \mid x_i])$, we may guide the search somewhat by looking only for those λ for which at the same time $cons((\lambda, x_i))$ and $cons(a_{i+1}(\lambda, x_i) \cup b_{i+1}(\lambda, x_i))$ Indeed, in many applications we may even be able to get a step further:

We often have a notion of consistent extension associated with attribute spaces (and their higher levels in $G(A)$). By this we understand a functional extension such that for every finite x, if x is consistent, then $\text{ext}(x)$ is a finite set $\text{ext}(x) = \{y_1, ..., y_k\}$ such that

(i) (y_i, x) is consistent for every i,

(ii) if M is any maximal consistent extension of x then there is j such that $M \cup \{y_i\}$ is consistent; i.e. the y_i form a consistent covering of x.

If such an extension notion is available, then the solution algorithm for $ax = bx$

starting at x_0 proceeds as follows: at each level x_i of approximation query

$$?: w([y_i \mid x_i])$$

for each $y_i \in ext(x_i)$. Augment the collection of windows by those (y_i, x_i) for which the answer is yes. If the query returns no, then set $x_{i+1} = (y_i, x_i)$ and iterate. If all queries return yes then backtrack to the previous level x_{i-1}.

The consistency notion which we introduced above clearly has the compactness property. From this it follows that every consistent set may be extended to a maximal consistent set(a fact that we used in the definition of covering extensions). It also follows that if $ax = bx$ has no (consistent) solution, then our proposed solution algorithms would terminate after finitely many steps with failure by exhaustion.

Acknowledgement. The present paper was essentially written in early 1988. It was then presented in lectures at various places and I wish to thank my friends for the occasions to discuss this approach to modelling. It is clear to anybody connected in the least way with the exciting field of knowledge-based modelling, that the present approach is but one of many, that our exposition lacks examples, and that we should perhaps have offered a long list of references. But I feel this would have been incongruous to such a short sketch.

In any case, the present sketch has served as a starting point to the thesis of *G.* Schwarzler at ETH who addressed the very-far-from-trivial problems of software engineering and computer algebra that arise when an actual programming environment for this discipline is to be established. His work contains the required references as well as some worked-out examples.

Computer Mathematics
Proc. of the Special Program at
Nankai Institute of Mathematics
January 1991 - June 1991

The Symmetry Groups of Computer Programs and Program Equivalence

J.R. Gabriel

M.H. Gabriel & Associates
680W.97th st.
Lemont, IL 60439

1. Introduction

This paper deals with the question of whether two different computer programs $x \longrightarrow y = f(x)$ and $x \longrightarrow y = g(x)$ implement the same mapping from x to y.

The methods discussed are new to the field of proofs about programs, and different from those of classical recursive function theory and Turing Automata. They originate with the work of Sophus Lie and Henri Poincarè at the end of the nineteenth century, and are variants of ideas quite well known in classical and quantum mechanics, as well as in function theory [13] [9] [8] [11].

It seems clear that a substantial capability in Computational Algebra will be needed to make them useful in practice, their possible usefulness comes from the fact that they are an alternative formal theory of program correctness to those already well known, which is able to exclude the difficulties in principle arising from theories including Turing Automata.

The theory leads to definitions of recursive procedures to enumerate the elements of finite sets which, although large, seem likely to be modular polynomial rings with coefficients from a finite field, usually the Galois Field of two elements.

A program is defined so as to include all real programs on finite machines, using finite resoureses, but excluding Turing Automata.

1.1. Definition A program is a single valued mapping $x > f(x)$ where x is an element of set S being the domain of the program. The set $\{y\} = f(S)$ is the range of the program.

1.2. Definition The programs $x > f(x)$ and $x > g(x)$ are equivalent iff

$$\forall x \in S, \quad f(x) = g(x)$$

1.3. Group Spaces

A set G of mappings of S onto itself, and obeying the group axioms is an important part of the theory. For finite sets S this always exists as all permutations of the elements of the domain, but in practice the domain may have a simple representation such as a Galois Field, a modular polynomial ring (the unsigned integers modulo 2^{32}) or a linear vector space (typical in linear algebra). These domains have simpler symmetry groups than the permutation group, and the requirements of a particular problem domain may simplify matters even further (for example in cryptology).

S is said to be a *Group Space* w.r.t G iff

1. $\forall x \in S$ and all $t \in G$ then $t(x) \in S$ so that S is invariant w.r.t G.

2. S is *transitive* w.r.t G. This idea needs further explanation.

If $x \in S$ the set $t(x) \forall t \in G$ is called the orbit of x w.r.t. G.
S is transitive w.r.t G iff $\forall x \in S$ the orbit of x w.r.t. G is itself.
If S is invariant and transitive w.r.t. G, S is said to be irreducible w.r.t. G.
If a domain is irreducible w.r.t. G, it is said to be a *Group Space* w.r.t. G.

1.4. Example

Let x_i for $i = 1, 2, 3, ..., n$ be S. Let G be the set of all permutations of the x_i among themselves. Then the exchanges $P_{ij} = x_i \longrightarrow x_j; x_j \longrightarrow x_i$ for given i generate S from x_i, and letting i vary, S is clearly transitive w.r.t. its permutation group. Because G is a group of permutations of the x_i among themselves, S is also invariant, so that S is irreducible and a group space.

2. Induced Representations

Let S be the domain of $f(x)$, and let it be a group space with group G, having elements γ.

If $y = f(x)$, and $y' = f(\gamma(x))$ then γ is said to *induce* the mapping $y > y'$ in the range R of $f(x)$, as x varies over S.

The mappings in R obey the group axioms, i.e. we may write $y' = f(\gamma(x)) = \Gamma(\gamma)f(x)$, and $\Gamma(\gamma_1\gamma_2) = \Gamma(\gamma_1)(\gamma_2)$, so that induced representation and G are related by a homomorphism, since products correspond to products.

It is also easy to prove that R is invariant and transitive w.r.t. the group of all Γ.

Clearly the homomorphism is determined by the program. Because both S and R are group spaces, the mapping between them is determined by the homomorphism and one pair $(x > y)$, i.e. not only does the program determine the homomorphism. but the homomorphism and one pair (x, y) determines the mapping performed by the program.

2.1. Corollary If two programs $f(x)$ and $g(x)$

* Have the same domain which is a group space.
* Induce the same representation of the group of the domain.
* And $f(x_0) = g(x_0)$ for a single x_0.

Then the two programs perform the same input output mapping and are equivalent according to the definition of this paper.

3. Input Equivalence Spaces

The next task is to study the connection between the program $y = f(x)$ and subgroups or representations of G. Let $E(f)$ be the subgroup of G corresponding to the identity in the representation induced from S onto R by the program , i.e. if G is the permutation group on S then $E(f)$ only permutes those elements of $S = \{\Sigma\}$ for which $f(x)$ is the same.

Let $S(f, y)$ be the set of all x such that $y = f(x)$. Then

1. $S(f, y)$ is invariant w.r.t.$E(f)$

2. If $x \in S(f, y)$ then the orbit of x is $S(f, y)$. Proof:-

Let x_0 be in $S(f, y)$. Let x_i be an element of $S(f, y)$ not in the orbit of x_0. Since $x_i \in S(f, y)$, $f(x_i) = y$, and the mapping $x_0 > x_i$ must be in $E(f)$, because if not G would contain an element mapping into the identity in R , not a member of $E(f)$. This is a contradiction, and so the assumption that x_i is not in the orbit os x_0 must have been false.

It follows that $S(f,y)$ is irreducible w.r.t. $E(f)$ and is a group space w.r.t. $E(f)$.

Since $E(f)$ corresponds to the identity in the induced representation $GE(f) = E(f)G$ is a set identity in G, i.e. $E(f)$ is an invariant subgroup of G. Thus elements of G transform left or right cosets of G w.r.t. $E(f)$ isomorphically with the representation of G induced representation.

The sets $S(f,y)$ are called ***input equivalence sets*** of $y = f(x)$, and $E(f)$ the ***input equivalence group***. Clearly the input equivalence group determines the input equivalence sets, and vice versa.

Since $f(x)$ is single valued, the sets $S(f,y)$ must be disjoint, and their direct sum must be S.

4. Examples

Consider the integer square root $y = sqrt(x)$, where S is the subset of the natural numbers bounded above by some natural number B,usually a power of 2. The result of the computation is the largest natural number y such that $y^2 \leq x$.

It is easily seen that the input equivalence sets are the closed intervals $y^2 \leq x < y^2 + 2y$. The cosets of $E(f)$ are $\gamma(y, y+1)E(f)$, where γ interchanges y with $y+1$ in the range of sqrt , and y^2 with $(y+1)^2$ in the domain.

As an exercise, the reader might consider the floating point square root in the domain $1.0 \leq x < 4.0$ for an ideal binary arithmetic. The input equivalence sets will each have two elements, one the exact square root of y, and the other the next floating point number above . The case for hexadecimal arithmetic is also interesting [4]

5. Invertibility and Measure

Note that $y = f(x)$ is an invertible mapping iff $E(f)$ contains only the identity. Invertible programs are called ***encryptions***. If a program $y = f(x)$ is not an encryption, then there are fewer points in the range of $f(x)$ than in its domain. This result showing decreasing measure may be useful in proof that recursion leads eventually to an encryption.

6. Composition of Programs

If the input equivalence sets of $y = f(x)$ and $z = g(y)$ are known, it seems reasonable to expect that the input equivalence sets of

$$z = h(x) = g(f(x))$$

can be determined. Such a determination opens the door in principle to calculation by symbolic algebra of the input equivalence sets of a program, given the input equivalence sets of the elementary statements of the programming language, and the text of the program.

Let $S(g, z)$ be an input equivalence set of $z = g(y)$, and $S(f, x)$ an input equivalence set of $y = f(x)$. Let S be the domain of $f(x)$ and let the domain of $g(y)$ be the range of $y = f(x)$.

Because input equivalence sets are disjoint,

$$S(h, z) = \Sigma_{y \in S(g,z)} S(f, y)$$

This is a simple result, although it may not be very simple to compute symbolically.

There is a corresponding result for subgroups. Since the domain of $g(y)$ is a group space w.r.t. the quotient group $G/E(f)$ where the domain of f is a group space w.r.t. G, and $E(f)$ is the input equivalence group of $x \longrightarrow f(x)$, then the input equivalence group of $y \longrightarrow g(y)$ is an invariant subgroup of $(G/E(f))/E(g)$. The results about measure can also be derived from transitivity of the space w.r.t. these quotient groups.

If instead of considering the composition of a program, one considers its decomposition into parts, it is fairly easily shown that there are lattices of subgroups analogous to the lattices of Denotational Semantics and so it seems there is a fairly close correspondence between the group theoretical concepts and those due to Scott, Strachey, and Stoy [12].

7. Iteration and Recursion

Suppose the range of $x \longrightarrow f(x)$ is a subset of its' domain. Suppose also that $e(x)$ has the same domain as $f(x)$ and a range of *TRUE/FALSE*.

Then one may consider the recursion

```
procedure h(x)
{
{while (NOT e(x))x = f(x);}
return x
}
```

There will be a chain of invariant subgroups

$$E(f) \subseteq E(f^2) \subseteq E(f^3) ... \subseteq E(f^n) ... \subset G$$

where the notation f^n stands for $f(f(f(...f(x)...)))$ recursed n times. The composition theorem shows that the range of $f^n(x)$ decreases monotonically until either $e(x)$ becomes *TRUE* or the range of the n'th iteration is such that $f(x)$ for any x from the range (i.e. the domain of the $n+1$'th iteration) is such that $f(x)$ for the domain is an encryption.

At this stage the subgroup chain is such that the input equivalence groups are all the identity, and $E(f^n) = E(f^{n+1}) = ...$ and the subgroup chain has attained a limit.

Because $f(x)$ is an encryption at this stage it is invertible, and generates a finite cyclic group. Call the range of $f^n(x)$ Δ.

If the input equivalence set $S(e, TRUE)$ of the exit condition, is a subset of the orbit of Δ w.r.t. the cyclic group generated by $f(x)$, then the iteration exists, otherwise there are values in the input domain for which the iteration will continue until ended by intervention from some external influence.

In practice, it seems that recursions are used either to obtain convergence (most typically in numeric work) or to search some solution space.

In the case where convergence is desired, the programmer usually has either an informal proof that exit will take place before the limit of $E(f^n)$) is reached, or defensive programming is practiced and an error exist is taken when some fixed number of the recursions have taken place without exit.

When a search is desired, it is best for $f(x)$ to be an encryption, since in this case the range does not diminish during iteration and more the domain can be covered in a given number of iterations. In this situation, the programmer usually has an informal proof that the orbit of the starting domain w.r.t. the cyclic group generated by $f(x)$ includes $S(e, TRUE)$, or again some defense is made against failure to terminate.

A more formal statement of the above can be made in terms of limit points or sets as is done for example in Denotational Semantics.

The symmetry theory seems also to be a generalization of the idea of loop invariants

8. Practical Considerations

It is not hard to show, essentially by constructing "successor" functions for th $S(f, y)$ that correspond to interchanges $P(i, i+1)$ generating the permutation grou of the domain, that composition can be posed as a recursive computation. Whethe the results of this paper lead to a uniform method for proof of equivalent as recursiv functions and proving them identical has still to be determined.

It seems fairly clear that the input equivalence sets, and corresponding symmetı groups can be determined almost by inspection for simple programming languages, but whether the "cure" of composition, is worse than the "disease" of straightforward attempts to prove equivalence is not known.

It does seem certain that the ease of any kind of proof is likely to depend on the complexity of the input domain of the program being examined.

It also seems likely that object oriented programming might be more amenable to attack by a combination of the means described here to the objects, and some earlier work on what amounts to derivation of properties of dataflow graphs from knowledge about the nodes and arcs [7] [1].

It is also clear that many programs are written and used in simulation for example, because parts of the system are well understood, but the system as a whole is not. In this case, proofs of isomorphism between parts of programs, and the subsystems of a real artifact being modeled, together with proof of equivalence between dataflow in the program and interaction between parts in the real world would lead to proof of truthful representation of the real world by the program without either the real world or the program being understood in its entirety.

The few examples of successful automated proofs about programs examined by the author – the integer square roots, for example, seem to have easily seen symmetry groups. An effort was made to use the Boyer Moore theorem prover [3] [2] to obtain symmetry groups for the two halves of an integer square root program, and then compose them, before most the theoretical results of this paper were known.

The disjointness theorem for input equivalence sets had not been recognized at the time, and some effort was spent on development of methods in the Boyer-Moore formalism for symbolic computation on possibly overlapping intervals.

9. Stronger Domain Symmetries

The domains considered here are not linear vector spaces, and so the powerful methods of Representations of Linear Groups are not appropriate in the very general theory. However, it seems that in practice domains often have richer structures than do simple sets. For example the "unsigned integers N bits long" are equivalent to a polynomial ring modulo x^N, and are a linear vector space. Similar things can probably be said of the floating point numbers. Some, and perhaps many, non numeric computations seem to be encryptions.

10. Conclusions

There are symmetry groups for programs, just as there are for classial and quantum dynamical systems, and for solutions of differential equations.

The theory excludes Turing Machines, and their associated difficulties, and shows how the behavior of a program may be described independently of the program text.

It remains to be seen whether the theory is useful in practice for more complex programs than those where proof of equivalence is possible by currently available means, and whether the composition theorem is a practical method of proof.

The theory is a substantial generalisation of the idea of a loop invariant, and the idea of an input equivalence sets has close connections with Inverse Error Analysis [14].

11. Other Open Questions

The section of this paper on iteration and recursion seems definitely open to further work along the lines in topology and differential geometry that have proved fruitful in Chaos Theory and Iterated Transformations [5],[6],[10].

The richness of the iterated mappings $x \longrightarrow f(x)$ in fractal geometry shows how complex the input/output relations of iterations on even very simple functions can be, and suggests that the many of the programs we actually write, have much simpler input/output relations and symmetry groups than would be the case in general. This raises interesting questions of taxonomy analogous to Chomsky's taxonomy of languages.

References

[1] *Proving the correctness of Digital Hardware Design,* Annual proceedings of the American Association for Artificial Intelligence, 1983.

[2] R. S. Boyer and J. S. Moore, *A Computational Logic* ACM, 1979.

[3] R. S. Boyer and J. S. Moore, *A Computational Logic Handbook,* Academic Press, New York, 1988.

[4] W. J. Cody and W. M. Waite, *Software for the Elementary Functions,* Prentice-Hall, Englewood Cliffs, New Jersey, 1980.

[5] Mitchell J. Feigenbaum, *Quantitative Universality for a Class of Nonlinear Transformations* J. Stat. Phys. 19, 25 1978.

[6] Mitchell J. Feigenbaum, *The Universal Metric Properties of Nonlinear Transformations* J. Stat. Phys. 21, 699 1979.

[7] J. R. Gabriel and R. Chapman, *Deriving the Properties of systems from Properties of Parts and Lists of Connections*, Argonne National Laboratory Report ANL-86-51, Argonne, Illinois, 1986.

[8] J. S. Lomont, *Applications of Finite Groups*, Academic Press NY, 1959.

[9] G. W. Mackey, *Induced Representation of Groups and Quantum Mechanics*, Publ. W. A. Benjamin NY 1968, also *Privately Circulated Lecture Notes* 1955. approx.

[10] N. Metropolis, M. L. Stein and P. R. Stein, *On Finite Limit Sets for Transformations on the Unit Interval*, J. Combinatorial Theory (A) 15, 25 1973.

[11] W. Miller, Jr., *Lie Theory and Special Functions*, Academic Press, NY, 1968.

[12] J. Stoy, *Denotational Semantics: The Scott Strachey Approach to Programming Language Theory*, MIT Press, Cambridge, Mass., 1977.

[13] H. Weyl *Gruppentheorie und Quantenmechanik*, Originally published in German in 1928, probably by J. von Springer Berlin, English translation by H. P. Robertson 1931, probably published by Methuen London, reprinted by Dover Press New York in the early 1950s.

[14] Wilkinson, J. H., *The Algebraic Eigenvalue Problem*, Publ. Oxford University Press, Oxford, U. K. 1965.

Computer Mathematics
Proc. of the Special Program at
Nankai Institute of Mathematics
January 1991 - June 1991

Computation with Rational Parametric Equations*

Shang-Ching Chou
Department of Computer Science
The Wichita State University, Wichita, KS 67260, USA

Xiao-Shan Gao
Institute of Systems Science, Academia Sinica
Beijing 100080, P.R. China

Zi-Ming Li
Research Institute for Symbolic Computation
Johannes Kepler University, A-4040 Linz, Austria

Abstract. Based on the characteristic method and the Gröbner basis method, we present algorithms for the following problems about the implicitization of the rational parametric equations. (1) To find a basis of the implicit prime ideal determined by a set of rational parametric equations. (2) To decide whether the parameters of a set of rational parametric equations are independent. If they are not, to reparameterize the parametric equations so that the new parametric equations have independent parameters. (3) To compute the inversion maps of parametric equations, and as a consequence, to give a method to decide whether a set of parametric equations is proper. In the case of algebraic curves, to find a proper reparameterization for a set of improper parametric equations. (4) To find a canonical representation for the image of a set of parametric equations, and as a consequence, to decide whether a set of parametric equations is normal. The algorithms presented in this paper have potential usage in geometric modeling, computer aided geometric design, etc.

* The work reported here was supported in part by Chinese NFS grant and by the NSF(USA) Grant CCR-917870.

1. Introduction

This paper is a summary of our study of various problems in the implicitization of rational parametric equations using Ritt-Wu's characteristic method and the Gröbner basis method [Gao & Chou,1990b,1990c,1990d], [Li, 1989]. New results are also included, particularly in Sections 6 and 8.

We may use two forms to represent some algebraic curves or surfaces: the *implicit form* and the *parametric form*. For example, the unit circle can be given by its implicit equation:

$$x^2 + y^2 - 1 = 0$$

or a set of parametric equations:

$$x = \frac{t^2 - 1}{t^2 + 1}, \; y = \frac{2t}{t^2 + 1} \tag{1.1}$$

Algebraic curves and surfaces are widely used in geometric modeling and it is recognized that both implicit and parametric representations for rational curves and surfaces have their inherent advantages: the parametric representation is best suited for generating points along a curve, whereas the implicit representation is most convenient for determining whether a given point lies on a specific curve. This motivates the search for a means of converting one representation to the other. For an algebraic curve or surface, converting its implicit form to the parametric form is called *parameterization*, and converting its parametric form to the implicit form is called *implicitization*. This paper addresses the problem of implicitization.

Implicitization and parameterization are classic topics in algebraic geometry. Only recently, due to their usage in geometric modeling, it has been studied extensively by researchers from the fields of computer science, see e.g. [Sederberg, 1984], [Arnon & Sederberg], [Buchberger, 1987], [Shannon & Sweedler, 1988], [Chuang & Hoffmann, 1989], [Li 1989], [Wu, 1989], [Hoffmann, 1990], [Chionh, 1990], [Manocha & Canny, 1990], [Ollivier, 1989], [Gao & Chou, 1990c, 1990d], and [Kalkbrener, 1990].

1.1. Implicit ideal and independent parameters

Each set of rational parametric equations determines a unique prime ideal which will be called the *implicit (prime) ideal* of the parametric equations. The difficulty of implicitization arises from the existence of base points. Based on the characteristic set method and the Gröbner basis, we present methods to find a basis of the implicit ideal for general *rational* parametric equations. Our methods fit for parametric equations with or without *base points*. A similar method using the Gröbner basis method has also been presented independently in [Kalkbrener, 1990].

In most cases, the parameters in a set of parametric equations are independent, i.e., the dimension of the implicit ideal equals the number of the parameters of the parametric equations. But there are parametric equations whose parameters might not be independent. Consider the following example:

$$x = \frac{u+v}{u-v}, y = \frac{2v^2+2u^2}{(u-v)^2}, z = \frac{2v^3+6u^2v}{(u-v)^3} \tag{1.2}$$

At first sight, one might think (1.2) represents a space surface. Actually, it represents a space curve, because if we let $t = \frac{u+v}{u-v}$, then the above parametric equations become

$$x = t,\ y = t^2+1,\ z = t^3-1.$$

For the above example, each point of the curve corresponds to infinitely many values of u and v. In practice, such "bad" parametric equations should be avoided. We give a method to decide whether the parameters of a set of rational parametric equations are independent. If the parameters of the parametric equations are not independent, we can reparameterize them so that the new parametric equations have independent parameters.

1.2. Inversion maps and proper parametric equations

The inversion problem is, given a point on the image of a set of parametric equations, to find a set of values for the parameters which corresponds to the given point. The inversion problem can be reduced to an equation solving problem [Buchberger, 1987]. In the following, we show that in certain cases, we can find a closed form solution to the inversion problem, i.e., we give a method to compute the inversion maps of the parametric equations. As a consequence of our method, we can decide whether the parametric equations are proper, i.e., whether the implicit curves or surfaces are not multiply traced by the parametric equations [Faux and Pratt 1979].

If the parametric equations are not proper, naturally we would ask whether we can reparameterize them so that the new parametric equations are proper. In general cases, the answer is negative. However, in the case of algebraic curves, the existence of a proper reparametrization for the original improper parametric equations is guaranteed by Lüroth's theorem [Walker, 1950]. A constructive proof of Lüroth's theorem actually provides an algorithm to construct a set of proper parametric equations. Recently Sederberg gave a method to find proper reparametrization for any set of improper parametric equations for algebraic curves [Sederberg 1986]. As an application of our method, we provide a new method to find a proper reparameterization for a set of improper parametric equations of an algebraic curve and our method does not need to randomly select sample points on the curve as Sederberg's algorithm does. In the case of algebraic surfaces, if the ground field K is the complex number field $\mathbf{C}$, then there always exists a proper reparameterization for the original improper parametric equations [Castelnuovo 1894]. However if the base field K is $\mathbf{Q}$ (the field of rational numbers) or $\mathbf{R}$ (the field of real numbers), this does not need to be the case [Seg 1951]. If the implicit variety determined by the parametric equations are of dimensio

> 2, then even for $K = \mathbf{C}$ there exist improper parametric equations that do not have a proper reparameterization [Artin & Mumford 1971].

1.3. The images and normal parametric equations

In the parametric representation for algebraic curves and surfaces, certain points on the curves or the surfaces may be missed by parametric equations. For example, the point (1,0) on the circle cannot be given by parametric equation (1.1). The following example shows that the missing part may have positive dimension.

$$(1.3) \qquad x = \frac{u^2 - v^2 + 1}{u^2 + v^2 - 1}, y = \frac{2vu}{u^2 + v^2 - 1}, z = \frac{2u}{u^2 + v^2 - 1}.$$

is a set of parametric equations of $x^2 + y^2 - z^2 = 1$. The missing part is $Zero(z, x^2 + y^2 - 1) \cup Zero(x + 1, z^2 - y^2) - Zero(z, y, x + 1)$.

To find the missing points, we need to compute the image of a set of parametric equations. In [Li, 1989], such a method is given based on an algorithm of projection of a quasi variety which is presented in [Wu, 1989]. The algorithm of projection of a quasi variety is in turn based on Ritt-Wu's decomposition algorithm [Wu, 1984, 1986]. In this paper, we extend Wu's projection algorithm to give a new method of quantifier elimination over an algebraic closed field.

Parametric equations of a curve or a surface are called *normal*, if all the points of the curve or the surface can be given by the parametric equations. In this paper, we give a method to decide whether parametric equations for a curve or a surface are normal. Also, some simple criteria for parametric equations to be normal are given. These criteria are very easy to use. Based on these criteria, we prove that polynomial parametric equations for a curve are always normal. As a conclusion, most of the rational curves used in CAD, such as the cubic Hermite curves, the Bezier curves, and the cubic B-spline curves [Pratt, 1986], are normal curves.

1.4. Experiment result and comparisons

We have implemented the algorithms described in this paper on a SUN3 workstation using Common Lisp. The central part of our implementation is an improved version of Ritt-Wu's decomposition algorithm [Chou & Gao, 1990a,b]. Compared with most papers mentioned above, our program deals with the most general case, not only for curves or space surfaces. Our program can do more things such as the computation of the image for a set of parametric equations and reparameterization of parametric equations such that the new parametric equations have independent parameters. Also, our program is efficient enough to solve non-trivial problems automatically. For more tetails of computation, see Section 8.

2. Preliminary on Rational Parametric Equations

Let K be a computable field of characteristic zero, e.g., $\mathbf{Q}$. We use $K[x_1,...,x_n]$ or $K[x]$ to denote the ring of polynomials in the indeterminates $x_1,...,x_n$. Unless explicitly mentioned otherwise, all polynomials in this paper are in $K[x]$. Let E be a *universal extension* of K, i.e., an algebraic closed extension of K which contains sufficiently many independent indeterminates over K. For a polynomial set PS, let

$$Zero(PS) = \{x = (x_1,...,x_n) \in E^n \mid \forall P \in PS, P(x) = 0\}.$$

For two polynomial sets PS and DS, we define

$$Zero(PS/DS) = Zero(PS) - \cup_{d \in DS} Zero(d).$$

Let $t_1,...,t_m$ be indeterminates in E which are independent over K. For polynomials P_1, ..., P_n, Q_1, ..., Q_n in $K[t_1,...,t_m]$ ($Q_i \neq 0$), we call

$$x_1 = \frac{P_1}{Q_1}, \ ..., \ x_n = \frac{P_n}{Q_n} \tag{2.1}$$

a set of (rational) parametric equations. We assume that not all P_i and Q_i are constants and $gcd(P_i, Q_i) = 1$. The maximum of the degrees of P_i and Q_j is called the *degree* of (2.1). The image of (2.1) in E^n is

$$IM(P,Q) = \{(x_1,...,x_n) \mid \exists t \in E^m(x_i = P_i(t)/Q_i(t))\}.$$

We have

Lemma 2.2. There is an algorithm to find polynomial sets $PS_1,...,PS_t$ and polynomials $d_1,...,d_t$ such that

$$IM(P,Q) = \cup_{i=1}^{t} Zero(PS_i/\{d_i\}). \tag{2.2.1}$$

Proof. It is obvious that $IM(P,Q) = \{(x_1,...,x_n) \mid \exists t \in E^m(Q_i(t)x_i - P_i(t) = 0 \wedge Q_i(t) \neq 0)\}$. Thus by the quantifier elimination methods for algebraically closed fields (see, e.g., [Wu, 1989]), we can find the PS_i and d_i such that (2.2.1) is correct. ∎

For more representation of the image, see [LI1]. We give a canonical representation for the image in Section 6 of this paper.

Definition 2.3. The implicit ideal of (2.1) is

$$I = \{F \in K[x] \mid F(P_1/Q_1,...,P_n/Q_n) \equiv 0\}.$$

Zero(I) is called the implicit variety of (2.1).

It is clear that I is a prime ideal whose dimension equals the transcendental degree of $K(P_1/Q_1, ..., P_n/Q_n)$ over K. The following result gives the relation between the image and the implicit variety of a set of parametric equations.

Theorem 2.4. Let V be the implicit variety of (2.1) and d the dimension of V. Then

(1) $IM(P,Q) \subset V$; and

(2) $V - IM(P,Q)$ is a quasi-variety with dimension less than d. Furthermore, we can find this quasi-variety.

Proof. (1) is clear from the definitions. By (2.2.1), $IM(P,Q) = \cup_{i=1}^{t} Zero(PS_i/\{d_i\})$. We can further assume that for each PS_i, $Ideal(PS_i)$ (the ideal generated by PS_i) is a prime ideal and d_i is not in $Ideal(PS_i)$. Let I be the implicit ideal of (2.1). Since $\eta = (P_1/Q_1, ..., P_n/Q_n) \in IM(P,Q)$, η must be in some components, say in $Zero(PS_1/\{d_1\})$. Note that η is a generic point of V and $Zero(PS_1) \subset V$. Then $Zero(PS_1) = V$ and $Ideal(PS_1) = I$. Hence $V - IM(P,Q) = Zero(I \cup \{d_1\}) - \cup_{i=2}^{t} Zero(PS_i/\{d_i\})$. Since d_1 is not in I, the dimension of $Zero(I \cup \{d_1\})$ is less than d. ∎

3. The Computation of the Implicit Ideal

For a set of rational parametric equations of the form (2.1), let

$$F_i = Q_i x_i - P_i, \quad D_i = Q_i z_i - 1, \quad i = 1, ..., n \tag{3.1}$$

where the z_i are new variables. Let

$$ID = Ideal(F_1, ..., F_n, D_1, ..., D_n) \tag{3.2}$$

i.e., the ideal generated by F_i and D_i in $K[t, x, z]$.

3.1. A Method Based on the Gröbner Basis Method

Theorem 3.3. We use the same notations in (3.1) and (3.2). The implicit ideal of (2.1) is $ID \cap K[x_1, ..., x_n]$.

Proof. By the proof of Theorem 2.4, the implicit ideal of (2.1) is

$$I = \{F \in K[x] \mid F(P_1/Q_1, ..., P_n/Q_n) \equiv 0\}.$$

For $B \in I$, replacing P_i/Q_i by $x_i - F_i/Q_i$ in $B(P_1/Q_1, ..., P_n/Q_n) = 0$ and clearing denominators, we have

$$(\prod_{i=1}^{n} Q_i^{k_i}) B(x_1, ..., x_n) = \sum_{j=1}^{n} C_j F_j \tag{3.3.1}$$

where $C_j \in K[x,t]$. Multiplying both sides of (3.3.1) by $G = \prod_{i=1}^{n} z_i^{k_i}$, we have

$$(3.3.2) \qquad (\prod_{i=1}^{n}(z_iQ_i)^{k_i})B(x_1,...,x_n) = \sum_{j=1}^{n} GC_jF_j$$

Since $D_i = Q_iz_i - 1$, (3.3.2) shows that $B(x_1,...,x_n)$ can be expressed as a linear combination of F_i and D_i. Therefore B is in $ID \cap K[x]$. Thus we have proved $I \subset ID \cap K[x]$. To prove the other direction, let $P \in ID \cap K[x]$. Then we have

$$P = \sum_{i=1}^{n} C_iF_i + \sum_{j=1}^{n} B_jD_j$$

Set $x_i = P_i/Q_i$, $z_i = 1/Q_i$ in the above equation. We have $P(P_1/Q_1,..,P_n/Q_n) = 0$, i.e., P is in I. This completes the proof. ∎

Using the following lemma and Theorem 3.3, we can compute a basis for the implicit ideal of (2.1).

Lemma 3.4 (Lemma 6.8 in [Buchberger, 1985]). Let GB be a Gröbner basis of an ideal $ID \subset K[x_1,...,x_n,y_1,...,y_k]$ in the pure lexicographic order $x_1 < ... < x_n < y_1 < ... < y_k$. Then $GB \cap K[x_1,...,x_n]$ is a Gröbner basis of $ID \cap K[x_1,...,x_n]$.

Example 3.5. For example (1.1), let

$$(3.5.1) \quad PS = \{(v-u)x+v+u, (v-u)^2y-2v^2-2u^2, (v-u)^3z+2v^3+6u^2v, (v-u)z_1-1\}$$

Note that we can omit $(u-v)^2z_2 - 1, (u-v)^3z_3 - 1$ because of the appearance of $(v-u)z_1 - 1$. Under the pure lexicographical order $x < y < z < u < v < z_1$, the Gröbner basis of *Ideal*(PS) is

$$(3.5.2) \quad \{y - x^2 - 1, z - x^3 + 1, (x+1)v + (-x+1)u, 2uyz_1 + x + 1, 2vz_1 + x - 1\}$$

By Theorem 3.3 and Lemma 3.4, a basis of the implicit ideal of (1.1) is $\{y - x^2 - 1, z - x^3 + 1\}$.

Remark. The inequation part $D_i = 0$ (which is equivalent to $Q_i \neq 0$) is essential for Theorem 3.3 to be true. In Example 3.5, let $PS' = PS - \{(v-u)z_1 - 1\}$ and the Gröbner basis *Ideal*(PS') be GB'. Then $GB' \cap K[x] = \emptyset$.

The following lemma will be used later.

Lemma 3.6. ID is a prime ideal of dimension m.

Proof. Similarly to the proof of Theorem 3.3, we have

$$ID = \{P \in K[t,x,z] \mid P(t_1,...,t_m,P_1/Q_1,...,P_n/Q_n,1/Q_1,...,1/Q_n) \equiv 0\}$$

i.e., ID is a prime ideal with $(t_1, ..., t_m, P_1/Q_1, ..., P_n/Q_n, 1/Q_1, ..., 1/Q_n)$ as a generic point. Therefore, the dimension of ID is m. ∎

3.2. A Method Based on the Characteristic Set Method

The method given in Section 3.1 is complete. But for complicated problems, the computation of the Gröbner bases is quite time consuming. In this section, we present a method based on Ritt-Wu's characteristic set method. The Char set method can be used to solve some problems which can not be solved by the Gröbner basis method due to time and space limit of the computers.

Using the same notations as introduced in (3.1), let $PS = \{F_1, \cdots, F_n\}$ and $DS = \{Q_1, \cdots, Q_n\}$. It is obvious that

$$IM(P,Q) = \{(x_1, ..., x_n) \mid \exists(\tau_1, \cdots, \tau_m) \in E^m \quad (\tau_1, \cdots, \tau_m, x_1, \cdots, x_n) \in Zero(PS/DS)\} \tag{3.7}$$

Note that under the variable order $t_1 < \cdots < t_m < x_1 < \cdots < x_n$, $PS = \{F_1, \cdots, F_n\}$ is an *irreducible ascending chain* in $K[t, x]$. Thus by Theorem A.3 (i.e., Theorem A.3 in the Appendix), $PD(PS)$ (for the definition of PD, see the Appendix) is a prime ideal of dimension m. Note that DS is the set of initials of the polynomials in PS. Then by (A.1.1) we have

$$Zero(PS/DS) = Zero(PD(PS)/DS). \tag{3.8}$$

By Theorem A.6 and (3.8), we can find an irreducible ascending chain ASC under a new variable order $x_1 < \cdots < x_n < t_1 < \cdots < t_m$ such that

$$Zero(PS/DS) = Zero(PD(ASC)/DS). \tag{3.9}$$

ASC has the same dimension m as PS. Hence ASC contains n polynomials. Then by changing the order of the variables properly, we can assume ASC to be

$$\begin{aligned} &A_1(x_1, \cdots, x_{d+1}), \cdots, A_{n-d}(x_1, \cdots, x_n), \\ &B_1(x_1, \cdots, x_n, t_1, \cdots, t_{s+1}), \cdots, B_{m-s}(x_1, \cdots, x_n, t_1, \cdots, t_m) \end{aligned} \tag{3.10}$$

where $d + s = m$. Note that the *parameter set* of ASC is $\{x_1, ..., x_d, t_1, ..., t_s\}$.

Theorem 3.11. The implicit variety of (2.1) is $V = Zero(PD(A_1, \cdots, A_{n-d}))$.

Proof. By Theorem 2.4 and Lemma 3.5, (2.1) defines a variety W of dimension d. By (3.7) and (3.9), it is clear that $IM(P,Q) \subset V$. Then $W \subset V$. By Theorem A.3, V is also of dimension d. Therefore $V = W$. ∎

Remark. An algorithm to compute a basis of $PD(A_1, \cdots, A_{n-d})$ can be found in [Chou, 1988].

Example 3.12. For example (1.2), let

$$PS = \{(v-u)x+v+u, (v-u)^2y-2v^2-2u^2, (v-u)^3z+2v^3+6u^2v\}$$
$$(3.12.1)\ DS = \{u-v\}$$

Under the pure lexicographical order $x < y < z < u < v < z_1$, by Theorem A.6 we have $Zero(PS/DS) = Zero(PD(ASC))$ where

$$(3.12.2) \qquad ASC = \{y-x^2-1, z-x^3+1, (x+1)v+(-x+1)u\}$$

By Theorem 3.11, the implicit ideal of (1.2) is $PD(y-x^2-1, z-x^3+1)$, i.e., the ideal generated by $y-x^2-1, z-x^3+1)$.

4. The Independent Parameters

We will use the notations introduced in (2.1), (3.1), and (3.2).

Definition 4.1. The parameters $t_1, ..., t_m$ of a set of parametric equations of the form (2.1) are called independent if the implicit ideal of (2.1) is of dimension m, or equivalently the transcendental degree of the field $K(P_1/Q_1, ..., P_n/Q_n)$ over K is m.

Lemma 4.2. Assume that we have constructed (3.10). Then the transcendental degree of $K' = K(P_1/Q_1, \cdots, P_n/Q_n)$ over K is $d = m - s > 0$.

Proof. By (2.1), the transcendental degree of $K' = K(P_1/Q_1, \cdots, P_n/Q_n)$ over K is the maximal number of the independent quantities $x_1 = P_1/Q_1, ..., x_n = P_n/Q_n$, and hence is d by (3.10). Since not all of P_i and Q_i are constants in K and $gcd(P_i, Q_i) = 1$, some x_i must depend on t effectively. Hence $d = m - s > 0$. ∎

By Definition 4.1, we have

Corollary 4.2.1. The parameters of (2.1) are independent iff $s = 0$.

Theorem 4.3. If the parameters of (2.1) are not independent, we can find a set of new parametric equations

$$(4.3.1) \qquad x_1 = P'_1/Q'_1, \cdots, x_n = P'_n/Q'_n$$

which has the same implicit variety as (2.1) and is with independent parameters.

Proof. By Theorem A.6, we can find (3.10) from (2.1). By Theorem A.5, we can assume that the initial I_i of B_i and the initial J_j of A_j in (3.10) are polynomials of the parameters of ASC, i.e., of $x_1, ..., x_d, t_1, ..., t_s$. Since Q_i is not in $PD(F_1, ..., F_n) = PD(ASC)$, by Lemma A.4 we can find a nonzero polynomial q_i of the parameters of ASC, i.e., $x_1, ..., x_d$ and $t_1, ..., t_s$, such that

$$(4.3.2) \qquad q_i \in Ideal(A_1, ..., A_{n-d}, B_1, ..., B_{m-s}, Q_i).$$

Let $M = \prod_{i=1}^{m-s} I_i \cdot \prod_{j=1}^{n} q_j$. Then M is a polynomial of $x_1, ..., x_d, t_1, ..., t_s$. Let $h_1, ..., h_s$ be integers such that when replacing t_i by h_i, $i = 1, ..., s$, M becomes a nonzero polynomial of $x_1, ..., x_d$. Let P'_i and Q'_i be the polynomials obtained from P_i and Q_i by replacing t_i by h_i, $i = 1, ..., s$. Now we have obtained (4.3.1). We still need to show that (4.3.1) satisfies the conditions in Theorem 4.3.

We assume that (4.3.1) defines a variety W and (3.1) defines a variety V. By the selection of h_i, it is clear that the image of (4.3.1) is contained in the image of (3.1). Therefore, we have $W \subset V$ by Definition 2.3. Since (3.4) is a characteristic set of $PD(F_1, ..., F_n)$, for each F_k, by (A.1.1) we have

$$JF_k = \sum_{i=1}^{n-d} G_i A_i + \sum_{j=1}^{m-s} C_j B_j$$

where J is a product of powers of the initials of A_i and B_j. Hence J is a polynomial of $x_1, ..., x_d$ and $t_1, ..., t_s$. Replacing t_i by h_i, $i = 1, ..., s$, in the above formula, we have

$$J'F'_k = \sum_{i=1}^{n-d} G'_i A_i + \sum_{j=1}^{m-s} C'_j B'_j \tag{4.3.3}$$

where $F'_k = Q'_k x_k - P'_k$. By the selection of h_i, $J' \neq 0$ is a polynomial of $x_1, ..., x_d$. By Theorem 3.6, $V = Zero(PD(A_1, ..., A_{n-d}))$ has a generic zero $x_0 = (x'_1, ..., x'_n)$ such that $x'_1, ..., x'_d$ are independent variables over K. Let B''_i be obtained from B'_i by replacing the x by x_0. Since the initial I'_i of B'_i is a polynomial of $x_1, ..., x_d$, B''_i is a nonzero polynomial of $t_{s+1}, ..., t_m$ with nonzero initials which are free of $t_{s+1}, ..., t_m$. Then $B''_1 = 0, ..., B''_{m-s} = 0$ have solutions for $t_{s+1}, ..., t_m$. Let such a set of solutions be $t'_{s+1}, ..., t'_m$. Now replacing x by x_0 and t_i by t'_i, $i = s+1, ..., m$ in (4.3.3), we have $J''F''_k = 0$. Since J' is a polynomial of $x_1, ..., x_d$, $J'' \neq 0$ by the selection of x_0. Thus $F''_k = Q''_k x'_k - P''_k = 0$. Since q_i is a polynomial of $x_1, ..., x_d$ and $t_1, ..., t_s$, by (4.3.2) $Q''_k \neq 0$. Hence $x_0 = (P''_1/Q''_1, ..., P''_n/Q''_n)$ is in $IM(P', Q') \subset W$. As x_0 is a generic zero of V, we have $V \subset W$. We have proved $V = W$. Since (4.3.1) defines a variety of dimension d, by Corollary 4.2.1, the parameters $t_{s+1}, ..., t_m$ of (4.3.1) are independent.∎

Example 4.5. For example (1.2), from (3.12.2), we have $d = 1, s = 1$; hence the parameters u and v are not independent. To reparameterize (1.1), by Theorem 4.4, we have to compute M. Since $prem(u - v, ASC) = 2u$, we have $M = 2(x+1)u$. Selecting a value of u, say 1, which does not make M zero, we get a new parametric equation

$$x = \frac{v+1}{1-v}, y = \frac{2v^2+2}{(1-v)^2}, z = \frac{2v^3+6v}{(1-v)^3}$$

which has the same implicit ideal as (1.1) and has an independent parameter v.

Theorem 4.3 can also be proved using the Gröbner basis method [Gao & Chou, 1990c].

5. Inversion Maps and Proper Parametric Equations

The inversion problem is, given a point $(a_1, ..., a_n)$ on the image of (2.1), to find a set of values $(\tau_1, ..., \tau_m)$ for the t_i such that $a_i = P_i(\tau_1, ..., \tau_m)/Q_i(\tau_1, ..., \tau_m)$, $i = 1, ..., n$.

Definition 5.1. *Inversion maps* for (2.1) are functions

$$t_1 = f_1(x_1, ..., x_n), ..., t_m = f_m(x_1, ..., x_n)$$

such that $x_i \equiv P_i(f_1, ..., f_m)/Q_i(f_1, ..., f_m)$ are true on the implicit variety V of (2.1) except a subset of V which has a lower dimension than that of V.

The inversion problem is closely related to whether a set of parametric equations is proper.

Definition 5.2. (2.1) is called *proper* if for each $(a_1, ..., a_n) \in IM(P, Q)$ there exists only one $(\tau_1, ..., \tau_m) \in E^m$ such that $a_i = P_i(\tau_1, ..., \tau_m)/Q_i(\tau_1, ..., \tau_m)$, $i = 1, ..., n$.

Now let us assume that the parameters $t_1, ..., t_m$ of (2.1) are independent, i.e., $s = 0$. Then (3.10) becomes

$$\begin{array}{l} A_1(x_1, \cdots, x_{m+1}) \\ \cdots \\ A_{n-m}(x_1, \cdots, x_n) \\ B_1(x_1, \cdots, x_n, t_1) \\ \cdots \\ B_m(x_1, \cdots, x_n, t_1, \cdots, t_m) \end{array} \tag{5.3}$$

Theorem 5.4. Using the same notations as above, we have

(a) $B_i(x, t_1, ..., t_i) = 0$ determine t_i $(i = 1, ..., m)$ as functions of $x_1, ..., x_n$ which are a set of inversion maps for (2.1).

(b) (2.1) is proper if and only if B_i are linear in t_i for $i = 1, ..., m$, and if this is the case, the inversion maps are

$$t_1 = U_1/I_1, ..., t_m = U_m/I_m$$

where the I_i and U_i are polynomials in $K[X]$.

Proof. Let M be the product of the initials of A_i, B_j. Then M is a polynomial of $x_1, ..., x_d$. Let $x' = (x'_1, ..., x'_n)$ be a zero on the implicit variety V of (2.1) such that $M(x') \neq 0$. Then similarly to the proof of Theorem 4.3, we can show that $B_i(x', t_1, ..., t_i) = 0$, $i = 1, ..., m$, determine a set of values $t' = (t'_1, ..., t'_m)$ for the t_i and $C_k(x', t', z_1, ..., z_k) = 0$, $k = 1, ..., n$, determine a set of values $z' = (z'_1, ..., z'_n)$ for the z_i.

Furthermore, (t', x', z') is a zero of ID (see (3.2)) which implies that $Q_i(t') \neq 0$. Thus $F_h(t', x') = P_h(t')x'_h - Q_h(t') = 0$, i.e., $x'_h = P_h(t')/Q_h(t')$. Noting that $Zero(M) \cap V$ has a lower dimension than that of V, we have proved (a).

To prove (b), first note that $B_i = 0$ $(i = 1, ..., m)$ are the relations between the x and $t_1, ..., t_i$ in ID' which have the lowest degree in t_i. Also, different solutions of $B_i = 0$ for the same x give the same value for the x_i. Since (5.3) is an irreducible asc chain, for a generic zero x' on the implicit variety V, $B_i(x', t_1, ..., t_i) = 0$, $i = 1, ..., m$, have no multiple roots for the t_i. Therefore a point $x \in IM(P, Q)$ corresponds to one set of values for t_i iff B_i are linear in t_i, $i = 1, ..., m$. Let $B_i = I_i t_i - U_i$ where I_i and U_i are in $K[x]$. Then the inversion maps are $t_i = U_i/I_i$, $i = 1, ..., m$. ∎

Theorem 5.4 gives a method to find the inversion maps and a method to decide whether the parametric equations are proper.

Remark. In the terminology of algebraic geometry, if (2.1) is proper, then the variety V defined by (2.1) is a rational variety, i.e., V is birational to E^m.

Theorem 5.5. If $m = 1$ and (2.1) is not proper, we can find a new parameter $s = f(t_1)/g(t_1)$ where f and g are in $K[t_1]$ such that the reparameterization of (2.1) in terms of s

(5.5.1) $$x_1 = \frac{F_1(s)}{G_1(s)}, \quad ..., \quad x_n = \frac{F_n(s)}{G_n(s)}$$

is proper.

Proof. Since $m = 1$, (2.1) defines a curve C. Let $K' = K(P_1/Q_1, ..., P_n/Q_n)$ be the rational field of C. Note that $P_1(t_1) - Q_1(t_1)\lambda = 0$ where $\lambda = P_1(t_1)/Q_1(t_1) \in K'$, then t_1 is algebraic over K'. Let $f(y) = a_r y^r + ... + a_0$ be an irreducible polynomial $K'[y]$ for which $f(t_1) = 0$. Then at least one of a_i/a_r, say $\eta = a_s/a_r$, is not in K. By a proof of Lüroth's theorem (p149, [Walker, 1950]), we have $K' = K(\eta)$. This means that $x_i = P_i/Q_i$ can be expressed as rational functions of η and η also can be expressed as a rational function of $x_i = P_i/Q_i$, i.e., there is a one to one correspondence between the values of the $x_i = P_i/Q_i$ and η. Therefore η is the new parameter we seek. Now the only problem is how to compute f.

By Theorem 5.4, we can find an inversion map $B_1(x_1, ..., x_n, t_1) = 0$ of the curve. Then B_1 is a relation between the x and t_1 with lowest degree in t_1 module the curve, in other words, $B'_1(y) = B_1(P_1/Q_1, ..., P_n/Q_n, y) = 0$ is a polynomial in $K'[y]$ with lowest degree in y such that $B'_1(t_1) = 0$, i.e., $B'_1(y)$ can be taken as $f(y)$. So s can be obtained as follows. If B_1 is linear in t_1 then (2.1) is already proper. We can take $s = t_1$. Otherwise let

$$B_1 = b_r t_1^r + \cdots + b_0$$

where the b_i are in $K[x]$. By (2.1), b_i can also be expressed as rational functions $a_i(t_1)$, $i = 1, ..., r$. At least one of a_i/a_r, say a_0/a_r, is not an element in K. Let $s = a_0/a_r$.

Eliminating t_1 from (2.1) and $a_r s - a_0$, we can get (5.5.1). Note that a_i comes from b_i by substituting x_j by P_j/Q_j, $j = 1, ..., n$, then $s = b_0/b_r$ is an inversion map of (5.5.1). ∎

Theorem 5.5 provides a new constructive proof for Lüroth's Theorem, i.e., we have

Corollary 5.5.1. Let $g_1(t), ..., g_r(t)$ be elements of $K(t)$, then we can find a $g(t) \in K(t)$ such that $K(g_1, ..., g_r) = K(g)$.

In [Gao & Chou, 1990c], we presented constructive proofs for Theorem 5.4 and Theorem 5.5 based on the Gröbner basis method. We use the following example to show our method.

Examples 5.6. Consider the parametric equations for a Bézier curve [Sederberg, 1986]:

$$(5.6.1) \qquad x = \frac{8s^6 - 12s^5 + 32s^3 + 24s^2 + 12s}{s^6 - 3s^5 + 3s^4 + 3s^2 + 3s + 1}$$
$$y = \frac{24s^5 + 54s^4 - 54s^3 - 54s^2 + 30s}{s^6 - 3s^5 + 3s^4 + 3s^2 + 3s + 1}$$

Let $HS = \{(s^6-3s^5+3s^4+3s^2+3s+1)x-(8s^6-12s^5+32s^3+24s^2+12s), (s^6-3s^5+3s^4+3s^2+3s+1)y-(24s^5+54s^4-54s^3-54s^2+30s), (s^6-3s^5+3s^4+3s^2+3s+1)z-1\}$. Under the variable order $y < s < z$, the Gröbner basis of $Ideal(HS)$ in $K(x)[s, y, z]$ is

$g_1 = 224y^3+(-2268x+7632)y^2+(-54x^2-1512x-480384)y+34263x^3-424224x^2+1200960x$

$g_2 = (15273x^2 + 1098792x - 9767808)s^2 + (7280y^2 + (-27006x - 125592)y - 174069x^2 + 598788x - 9767808)s - 7280y^2 + (27006x + 125592)y + 189342x^2 + 500004x$

$g_3 = (488736x + 39071232)z + (33488y^2 + (-95718x + 1701432)y - 712134x^2 + 9970488x - 34187328)s + 27888y^2 + (-81210x + 1297128)y - 584109x^2 + 8885196x - 39071232$

Since (5.6.1) defines a plane curve, by Theorem 3.3 and Theorem 5.5, (5.6.1) is a set of improper parametric equations for the curve $g_1 = 0$. To find a set of proper parametric equations for $g_1 = 0$, by Theorem 5.5, we select a new parameter

$$(5.6.2) \qquad t_1 = \frac{(7280y^2 + (-27006x - 125592)y - 174069x^2 + 598788x - 9767808)}{(15273x^2 + 1098792x - 9767808)} = \frac{s^2+1}{1-s}$$

Eliminating s from (5.6.2) and (5.7.1), we have

$$(5.6.3) \qquad x = \frac{8t_1^3 + 12t_1^2 - 36t_1 + 16}{t_1^3 + 3t_1^2 - 3t_1}, y = \frac{-24t_1^2 + 78t_1 - 54}{t_1^3 + 3t_1^2 - 3t_1}$$

By Theorem 5.5, we can easily check that (5.6.3) is a set of proper parametric equations of $g_1 = 0$ with an inversion map (5.6.2).

6. The Computation of the Image

6.1. A Quantifier Elimination Method

By Lemma 2.2, to compute the image, we need to eliminate existence quantifiers, or to find the projection of a quasi variety. In [Wu, 1989], such a method has been provided. Wu's projection method is actually a special case of the so called *quantifier elimination method over an algebraic closed field* which has been given in [Tarski, 1951]. Also see [Heintz, 1983] for the complexity analysis. However, due to the well developed techniques for Ritt-Wu's decomposition algorithm, Wu's projection method is efficient.

Let PS and DS be polynomial sets in $K[x_1, ..., x_n, y_1, ..., y_m]$. The projection of $Z = Zero(PS/DS)$ with $y_1, ..., y_m$ is

$$Proj\ Z = \{(a_1, ..., a_n) \in E^n \mid \exists (b_1, ..., b_m) \in E^m, (a_1, ..., a_n b_1, ..., b_m) \in Z\}$$

where E is defined in Section 2.

Theorem 6.1. [WU2] With the notations defined above, we can find polynomial sets PS_i and polynomials G_i, $i = 1, ..., k$ such that

$$Proj\ Zero(PS/DS) = \cup_{i=1}^{k} Zero(PS_i/\{G_i\})$$

In the language of quantifiers, Theorem 6.1 is equivalent to

Corollary 6.1.1. For polynomials $P_1, ..., P_r, Q$ in $K[X, Y]$, we can find polynomials $F_{i,j}, G_i, i = 1, ..., t, j = 1, ..., d_i$, such that

$$\exists y_1 ... \exists y_m (P_1 = 0 \wedge \cdots \wedge P_r = 0 \wedge Q \neq 0)$$

is true iff the x satisfy one of the following restrictions

$$(F_{i,1} = 0 \wedge \cdots \wedge F_{i,d_i} = 0 \wedge G_i \neq 0), \quad i = 1, ..., t$$

It is a known result in logic that, if an existential quantifier can be eliminated, then there is a quantifier elimination theory for the theory of algebraic closed field (p. 83 [Shoenfield, 1967]). In the following, we will give a direct proof of this result.

Definition 6.2. A *formula of equation type* can be defined inductively as follows:

i) a polynomial equation or a polynomial inequation is a formula;

ii) if f, g are formulas, then $\neg f$, $f \wedge g$, and $f \vee g$ are formulas.

iii) if f is a formula, then $\exists x f$ and $\forall x f$ are formulas.

We have the following quantifier elimination algorithm over an algebraic closed field.

Theorem 6.3. Let f be a formula of equation type then we can find a formula of equation type g free of quantifiers such that f is true over an algebraic closed field iff g is true over an algebraic closed field.

Proof. It is easy to show that a formula of equation type can be expressed as the following canonical form:

$$Q_1y_1 \cdots Q_my_m(\vee_{i=1}^{r} \wedge_{j=1}^{t_i} P_{i,j}) \tag{6.3.1}$$

where Q_k is a quantifier and each $P_{i,j}$ is a polynomial equation or a polynomial inequation. A formula like (6.3.1) with the quantifiers deleted is called a *disjunctive formula.* If $Q_m = \exists$, by the following statement

$$\exists x(f \vee g) \text{ iff } (\exists x f) \vee (\exists x g)$$

(6.3.1) is equivalent to

$$Q_1y_1 \cdots Q_{m-1}y_{m-1}[\vee_{i=1}^{r} \exists y_m(\wedge_{j=1}^{t_i} P_{i,j})] \tag{6.3.2}$$

Note that for polynomials P and Q, $P \neq 0 \wedge Q \neq 0$ iff $PQ \neq 0$, then we can use Corollary 6.1.1 to eliminate y_m and obtain a new formula like (6.1.1) but with $m-1$ quantifiers. If $Q_m = \forall$, (6.1.1) is equivalent to

$$Q_1y_1 \cdots Q_{m-1}y_{m-1}\neg\exists y_m \neg(\vee_{i=1}^{r} \wedge_{j=1}^{t_i} P_{i,j}) \tag{6.3.3}$$

Since the negation of a disjunctive formula can be transformed to a equivalent disjunctive formula, (6.3.3) can be reduced to the following form

$$Q_1y_1 \cdots Q_{m-1}y_{m-1}\neg\exists y_m f$$

where f is a disjunctive formula. Now we eliminate y_m by using Corollary 6.1.1 and obtain a formula

$$Q_1y_1 \cdots Q_{m-1}y_{m-1}\neg g$$

where g is a disjunctive formula. Transforming $\neg g$ to a disjunctive formula, we obtain a formula like (6.1.1) but with $m-1$ quantifiers. Repeating the above process, we can eliminate the variables $y_{m-1}, \ldots, y_1$. ∎

6.2. A Canonical Representation for the Images

We have the following canonical representation for the image of a set of parametric equations.

Theorem 6.4. Let

$$x_1 = \frac{P_1}{Q_1}, \ldots, x_n = \frac{P_n}{Q_n} \tag{6.4.1}$$

be a set of parametric equations, then we can find irreducible asc chains ASC, ASC_i and polynomial sets DS_i, $i = 1, ..., k$, such that

$$IM(P,Q) = Zero(PD(ASC)) - \cup_{i=1}^{k} Zero(ASC_i/J_i \cup DS_i).$$

where J_i is the initial set of ASC_i. We also have (a) $PD(ASC)$ is the implicit ideal of (6.4.1); (b) $Zero(ASC_i/J_i \cup DS_i) \subset Zero(PD(ASC)$.

Proof. Since $IM(P,Q) = \{(x_1, ..., x_n) \mid \exists t \in E^m(Q_i(t)x_i - P_i(t) = 0 \wedge Q_i(t) \neq 0)\}$, by Theorem 6.1, we can find polynomial sets PS_i and DS_i in $K[x]$ such that

$$IM(P,Q) = \cup Zero(PS_i/DS_i)$$

By Ritt-Wu's decomposition algorithm, we further assume

$$IM(P,Q) = \cup Zero(ASC_i/DS_i \cup J_i) \tag{6.4.2}$$

where ASC_i are irreducible asc chains and J_i are the initial sets of ASC_i. Let us assume that ASC_1 is the asc chain which contains the least number of polynomials among the ASC_i. Then $Zero(ASC_1/DS_1 \cup J_1)$ is a component with maximal dimension in decomposition (6.4.2). By Lemma 2.4, the implicit variety of (6.4.1) is $Zero(PD(ASC_1))$. Then $Zero(ASC_i/DS_i \cup J_i) \subset IM(P,Q) \subset Zero(PD(ASC_1))$ for all i.

We have to transform (6.4.2) to the desired form. By the remainder formula (A.1.1), $Zero(ASC_1/J_1) = Zero(PD(ASC_1)/J_1)$. Then

$$\begin{aligned} IM(P,Q) &= Zero(PD(ASC_1)/DS_1 \cup J_1) \bigcup \cup Zero(ASC_i/DS_i \cup J_i) \\ &= (Zero(PD(ASC_1)) - Zero(\{D\})) \bigcup \cup Zero(ASC_i/DS_i \cup J_i) \end{aligned}$$

where $D = \prod_{I \in J_1} I \cdot \prod_{d \in DS} d$. Since $Zero(ASC_i/DS_i \cup J_i) \subset Zero(PD(ASC_1))$, we also have $IM(P,Q) = Zero(PD(ASC_1)) - W$ where

$$\begin{aligned} W &= Zero(\{D\}) - \cup_i Zero(ASC_i/DS_i \cup J_i)) \\ &= \cap_i (Zero(\{D\}) - Zero(ASC_i/DS_i \cup J_i)) \\ &= \cap_i [(Zero(\{D\}) - Zero(ASC_i)) \cup Zero(DS_i \cup J_i \cup \{D\} \cup ASC_i)] \\ &= \cap_i [\bigcup_{a \in ASC_i} Zero(\{D\}/\{a\}) \cup Zero(DS_i \cup J_i \cup \{D\} \cup ASC_i)] \end{aligned}$$

Using the following formula

$$Zero(PS_1/DS_1) \cap Zero(PS_2/DS_2) = Zero(PS_1 \cup PS_2/DS_1 \cup DS_2)$$

W can be written as $\cup_j Zero(RS_j/TS_j)$ for finite polynomial sets RS_j and TS_j. Using Ritt-Wu's decomposition again, we obtain the desired formula. ∎

A finite basis of $PD(ASC)$ can be computed; ref. e.g. [Chou, 1988].

Example 6.5. To compute the image of (1.3), let $PS = \{(u^2+v^2-1)x - u^2+v^2-1, (u^2+v^2-1)y - 2uv, (u^2+v^2-1)z - 2u\}$, $DS = \{u^2+v^2-1\}$. Using our program based on Theorem 6.1, we have

$$Proj\ Zero(PS/DS) = Zero(z^2-y^2-x^2+1/z(x+1)(y^2+x^2-1)) \cup Zero(z,y,x+1)$$

To find a canonical form for the image, using Theorem 6.4, we have

$$\begin{aligned} Proj Zero(PS/DS) =& Zero(z^2-y^2-x^2+1) \\ & - (Zero(z, y^2+x^2-1/x+1) \cup Zero(x+1, z^2-y^2/y)). \end{aligned}$$

7. The Normal Parametric Equations

Generally speaking, the image of a set of parametric equations is a quasi algebraic set. In this section, we discuss when the image of a set of parametric equations is an algebraic set.

Definition 7.1. (2.1) is called a set of *normal parametric equations* if $IM(P,Q)$ is an irreducible variety.

Theorem 7.2. We can decide in a finite number of steps whether parametric equations of the form (2.1) are normal parametric equations.

Proof. By Theorem 6.4, we can find a finite polynomial set PS such that the ideal generated by PS is a prime ideal and a quasi variety $W = \cup_{i=1}^{l} Zero(PS_i/DS_i)$ such that $IM(P,Q) = Zero(PS) - W$. Then (1.2) is normal if and only if $IM(P,Q) = Zero(PS)$, or equivalently $Zero(PS)$ and W have no common points. Without loss of generality, we only need to show how to decide whether $W' = Zero(PS) \cap Zero(PS_1, DS_1)$ is empty. Noting that $W' = Zero(PS \cup PS_1, DS_1)$, we can decide whether W' is empty using Ritt-Wu's decomposition algorithm [Wu, 1984]. ∎

The method in Theorem 7.2, though complete, usually needs expensive computations. In what follows, we give some simple criteria for normal parameterization which can be used without any computational costs.

Lemma 7.3. If the image $IM(P,Q)$ of (2.1) is an algebraic set, (2.1) are normal parametric equations.

Proof. Let $IM(P,Q) = Zero(PS)$, and let (2.1) be parameter equations of the irreducible variety V. By (2) of Definition 2.3, a generic point of V is in $Zero(PS)$. Thus $V \subset Zero(PS)$. By (1) of Definition 2.3, we have $Zero(PS) \subset V$, and hence $Zero(PS) = V$. ∎

Theorem 7.4. Let $y_1 = u_1(t)/v_1(t), ..., y_n = u_n(t)/v_n(t)$ be parametric equations of an algebraic curve. If $degree(u_i) > degree(v_i)$ for some i, they are normal parametric equations.

Proof. Let $RS = \{r_1, ..., r_h\}$ $(r_i \in K[y])$ be the resultant system of $h_1(t) = u_1(t) - v_1(t)y_1, ..., h_n(t) = u_n(t) - v_n(t)y_n$ for variable t (see p.158 Vol.1, [Hodge & Pedoe, 1952]. Then for any $y_0 = (y_{0,1}, ..., y_{0,n}) \in E^n$, $r_i(y_0) = 0, i = 1, ..., h$ if and only if $h'_1 = u_1 - v_1 y_{0,1} = 0$, ..., $h'_n = u_n - v_n y_{0,n} = 0$ have common solutions for t or the leading coefficients of $h'_1(t), ..., h'_n(t)$ all vanish. The later case is impossible, because there is an i_0 such that $degree(u_{i_0}(t)) > degree(v_{i_0}(t))$; hence the leading coefficient of $h'_{i_0}(t)$ is a nonzero number in K. Therefore, $r_i(y_0) = 0, i = 1, ..., h$ if and only if $h'_1(t) = 0, ..., h'_n(t) = 0$ have a common solution t_0. We have that $v_i(t_0) \neq 0$ for all i, for otherwise $u_i(t_0) = v_i(t_0)y_{0,i} = 0$. Therefore $u_i(t)$ and $v_i(t)$ have common solutions which contradicts the fact $gcd(u_i, v_i) = 1$. Thus $y_0 = (u_0(t_0)/v_0(t_0), ..., u_n(t_0)/v_n(t_0))$ is in the image $IM(u, v)$ of the parametric equations, i.e., $Zero(RS) \subset IM(u, v)$. It is easy to show that $IM(u, v) \subset Zero(RS)$. We have proved that $IM(u, v) = Zero(RS)$. By Lemma 2.7, the theorem has been proved. ∎

Corollary 7.4.1. A set of polynomial parametric equations of a curve is normal.

If a set of parametric equations (2.1) is not normal, then naturally we shall ask whether we can find a set of normal parametric equations which has the same implicit variety as (2.1). This problem is unsolved in general. But if the implicit variety of the parametric equation is a conic, then we have a solution to the above problem [Gao &Chou, 1990b]. As an example, we have

Example 7.5. The image of the following parametric equations

$$x = \frac{t^4 - 4t^2 + 1}{t^4 + 1}, \quad y = \frac{2\sqrt{2}(-t^3 + t)}{t^4 + 1}. \tag{7.5.1}$$

is $Zero(x^2 + y^2 - 1)$, i.e., (7.5.1) is a set of normal parametric equations for the unit circle. We have proved in [Gao & Chou, 1990b] that there exist no real coefficients quadratic parametric equations for the unit circle.

8. Special Techniques for Parametric Equations of Space Surfaces

Both Theorem 3.3 and Theorem 3.11 provide complete methods to find a basis for the implicit prime ideal of a set of parametric equations. But to compute the Gröbner basis or to obtain Ritt-Wu's decomposition is quite difficult in general. The complexity to compute the Gröbner basis is double exponential. The complexity of computing a characteristic set is single exponential [Gallo & Mishra, 1990]. But for some parametric equations, we may use their special property to develop efficient algorithms. In what follows, we consider such a partial algorithm.

Consider a set of parametric equations for a space surface

$$(8.1) \qquad x = \frac{P_1(t,s)}{Q_1(t,s)}\ y = \frac{P_2(t,s)}{Q_2(t,s)}\ z = \frac{P_3(t,s)}{Q_3(t,s)}$$

where P_i and Q_i are polynomials in $K[t,s]$. Let

$$(8.2) \qquad F_1 = Q_1x - P_1, F_2 = Q_2y - P_2, F_3 = Q_3z - P_3$$

The essential step in the implicitization is the triangulation of F_1, F_2, and F_3 by polynomial remainder sequences. We may use two tricks. First, we can use Collins' method or a similar result in the multi-polynomial case [Li, 1987] to remove extraneous factors produced in the computation. Secondly, we may choose a "good" way to do pseudo divisions. The purpose is to keep the degrees of the polynomials occurred in the computation as low as possible. We shall explain this in the following examples. We may also use the weak ascending instead of ascending chain to reduce the size of the polynomials. A set of polynomials $ASC = \{A_1, ..., A_p\}$ is called a *weak ascending chain*, if $class(A_i) < class(A_j)$ for $i < j$ and for each i the pseudo remainder of the initial of A_i with ASC is not zero [Chou & Gao, 1990a]. After the triangulation, we only consider a class of special case.

Theorem 8.3. Suppose we obtain the following triangular form from F_1, F_2 and F_3 under the variable order $x < y < z < s < t$

$$(8.3.1) \qquad g_1(x,y,z), g_2 = I_2(x,y,z)s - U_2(x,y,z), g_3 = I_3(x,y,z)t - U_3(x,y,z).$$

If $I_2(\eta)I_3(\eta) \not\equiv 0$ where $\eta = (\frac{P_1}{Q_1}, \frac{P_2}{Q_2}, \frac{P_3}{Q_3})$, then we have

(1) One irreducible factor of g_1, say f_1 (satisfying $f_1(\frac{P_1}{Q_1}, \frac{P_2}{Q_2}, \frac{P_3}{Q_3}) \equiv 0$), is the implicit equation of (8.1).

(2) Parametric equations (8.1) are proper and a set of inversion map is $t = \frac{U_2}{I_2}, s = \frac{U_3}{I_3}$.

Proof. Let $ID = Ideal(F_1, F_2, F_3, D_1, D_2, D_3)$ where $D_i = Q_iz_i - 1$, $i = 1, ..., 3$. By Lemma 3.3, ID is a prime ideal in $A = K[x,y,z,t,s]$ and the implicit ideal of (8.1) is $ID \cap K[x,y,z]$. It is easy to show that for $P \in A$,

$$(8.3.2) \qquad P \in ID \text{ iff } prem(P, F_3, F_2, F_1) = 0$$

Let $g_1 = f_1^{n_1} \cdots f_r^{n_r}$ be an irreducible factorization of g, then by (8.3.2) we can find a factor of g_1, say f_1 which is in ID. We shall show that

$$(8.3.3) \qquad f_1, g_2, g_3, D_1, D_2, D_3$$

consist of a (weak) characteristic set ofID under the variable order $x < y < z < t < s < z_1 < z_2 < z_3$. Since $I_2(\eta)I_3(\eta) \not\equiv 0$, by (8.3.2) I_2 and I_3 are not in ID. Note that

Q_i are also not in ID, then (8.3.3) is a weak ascending chain. The prime ideal with (8.3.3) as a characteristic set has the same dimension as ID. Therefore, (8.3.3) is a characteristic set of ID and Theorem 8.3 comes from this property and Theorem 5.4 immediately. ∎

Consider the following examples from [Hoffmann, 1990].

Example 1. (quardratic)

$$f_1 = z + (-9s + 15)t - 12s - 34$$
$$f_2 = y + (s - 8)t - 6s^2 - 7$$
$$f_3 = x - 3t^2 + (-s + 5)t - 4s^2 + 2s - 4$$

Example 2. (cubic)

$$f_1 = z - 2t^3 + (5s - 1)t + s^3$$
$$f_2 = y + (-s^2 + 3)t - 1$$
$$f_3 = x + t^3 - 3st - s^3 - s$$

Example 3. (bi-cubic)

$$f_1 = z - -3s(s^2 - 5s + 5)t^3 - 3(s^3 + 6s^2 - 9s + 1)t^2 + t(6s^3 + 9s^2 - 18s + 3) - 3s(s - 1)$$
$$f_2 = y - 3s(s - 1)^2 + t^3 + 3t$$
$$f_3 = x - 3t(t - 1)^2 + (s - 1)^3 + 3s$$

The computation of Example 1 is easy. To compute Example 2, we first form the pseudo remainders $f_4 = prem(f_1, f_2)$, $f_5 = prem(f_1, f_3)$, $f_6 = prem(f_2, f_3)$. Then f_6 and f_4 are polynomials of x, y, z, s such that $deg_s(f_6) = 5$ and $deg_s(f_4) = 9$. We compute the subresultant remainder sequence of f_4 and f_6 to obtain a triangular form. For Example 3, we first form the pseudo remainders $f_4 = prem(f_1, f_2)$, $f_5 = prem(f_3, f_2)$, $f_6 = prem(f_4, f_5)$, $f_7 = prem(f_2, f_5)$, $f_8 = prem(f_6, f_7)$, $f_9 = prem(f_5, f_7)$. Then f_8 and f_9 are polynomials of x, y, z, s such that $deg_s(f_8) = 9$ and $deg_s(f_9) = 9$. We compute the subresultant remainder sequence of f_8 and f_9 to obtain the triangular form. We have

(1) For Example 1, g_1 is an irreducible polynomial of degree 4 with 32 terms; g_2 is a polynomial of degree 1 in s with 38 terms; g_3 is a polynomial of degree 1 in t with 5 terms.

(2) For Example 2, g_1 is an irreducible polynomial of degree 9 with 176 terms; g_2 is a polynomial of degree 1 in s with 468 terms; g_3 is a polynomial of degree 1 in t with 6 terms.

(3) For Example 3, g_1 is an irreducible polynomial of degree 18 with 715 terms; g_2 is a polynomial of degree 1 in s with 1644 terms; g_3 is a polynomial of degree 1 in t with 12 terms.

(4) The running time for the three examples in a Symbolics-3600 are: 3 (secs), 397 (secs), 28315 (secs) respectively.

It is easy to see that all the three examples satisfy the conditions in Theorem 8.3. Therfore, it is easy to obtain the implicit equation and the inversion maps of the examples.

In our implementation of the algorithms of implicitization, we use the special techniques introduced in this section to treat parametric equations of space surfaces.

9. Conclusions

The main results of this paper can be summarized as follows.

For a set of rational parametric equations of the form (2.1),

(a) We can find a basis for the implicit ideal of (2.1).

(b) We can decide whether the parameters $t_1, ..., t_m$ are independent, and if not, reparameterize (2.1) such that the parameters of the new parametric equations are independent.

(c) If the parameters of (2.1) are independent, we can construct a set of polynomial equations

$$B_1(x_1, ..., x_n, t_1) = 0, ..., B_m(x_1, ..., x_n, t_1, ..., t_m) = 0.$$

The solution of the t_i in terms of the x_i are the inversion maps of (2.1), and (2.1) is proper iff the B_i are linear in t_i, $i = 1, ..., m$.

(d) If $m = 1$ and (2.1) is not proper, we can reparameterize (2.1) such that the new parametric equations are proper.

(e) We can express the image of a set of parametric equations as an irreducible variety with some lower dimension parts removed.

(d) We can find a canonical representation for the image of a set of parametric equations, and as a sequence, decide whether a set of parametric equations is normal. We can also find a normal reparametrization for a set of non-normal parametric equations of conics.

The general case of (d), i.e., to decide weather the implicit variety of (2.1) is rational (or equivalently, birational to E^k for some k), and if it is, to find a set of proper reparameterization for (2.1), is still open. For the case $m = 2$, see [Gao & Chou, 1990b] for some partial results.

10. Appendix. Some Results about Ritt-Wu's Decomposition Algorithm

A detailed description of Ritt-Wu's decomposition algorithm can be found in [Wu, 1984]. The implementation of the algorithms in this paper is based on a new version of the decomposition algorithm in [Chou & Gao, 1990b].

Let P be a polynomial. The *class* of P, denoted by $class(P)$, is the largest p such that some x_p actually occurs in P. If $P \in K$, $class(P) = 0$. Let a polynomial P be of class $p > 0$. The coefficient of the highest power of x_p in P considered as a polynomial of x_p is called the *initial* of P. For polynomials P and G with $class(P) > 0$, let $prem(G; P)$ be the *pseudo remainder* of G wrpt P.

A sequence of polynomials $ASC = A_1, ..., A_p$ is said to be an *ascending (*ab. *asc) chain*, if either $p = 1$ and $A_1 \neq 0$ or $0 < class(A_i) < class(A_j)$ for $1 \leq i < j$ and A_k is of higher degree than A_m for $m > k$ in x_{n_k} where $n_k = class(A_k)$.

For an asc chain $ASC = A_1, ..., A_p$ with $class(A_1) > 0$, the pseudo remainder of a polynomial G wrpt ASC is defined inductively as

$$prem(G; ASC) = prem(prem(G; A_p); A_1, ..., A_{p-1}).$$

Let $R = prem(G; ASC)$, then from the computation procedure of the pseudo division procedure, we have the following important *remainder formula:*

$$(A.1) \qquad JG = B_1A_1 + \cdots + B_pA_p + R$$

where J is a product of powers of the initials of the polynomials in ASC and the B_i are polynomials. For an asc chain ASC, we define

$$PD(ASC) = \{g \mid prem(g, ASC) = 0\}$$

By (A.1), a zero of ASC which does not annul the initials of the polynomials in ASC is a zero of $PD(ASC)$. More precisely, we have

$$(A.1.1) \qquad Zero(PD(ASC)) = Zero(ASC/J) \bigcup \cup_{d \in J} Zero(PD(ASC) \cup \{d\})$$

where J is the set of initials of the polynomials in ASC.

For an asc chain $ASC = A_1, ..., A_p$, we rename the variables. If A_i is of class m_i, we rename x_{m_i} as y_i, other variables are renamed as $u_1, ..., u_q$, where $q = n - p$. The variables $u_1, ..., u_q$ are called a *parameter set* of ASC. ASC is said to be an *irreducible ascending chain* if A_1 is irreducible, and for each $i \leq p$ A_i is an irreducible polynomial of y_i in $K_{i-1}[y_i]$ where $K_{i-1} = K(u)[y_1, ..., y_{i-1}]/D$ where D is the ideal generated by $(A_1, ..., A_{i-1})$ in $K(u)[y_1, ..., y_{i-1}]$.

Definition A.2. The dimension of an irreducible ascending chain $ASC = A_1, ..., A_p$ is defined to be $DIM(ASC) = n - p$.

Theorem A.3. ([Wu, 1984]) If ASC is an irreducible ascending chain then $PD(ASC)$ is a prime ideal with dimension $DIM(ASC)$.

Lemma A.4. ([Wu, 1984]) Let ASC be an irreducible asc chain with parameters $u_1, ..., u_q$. If Q is a polynomial not in $PD(ASC)$, then we can find a nonzero polynomial P in the u alone such that $P \in Ideal(ASC, Q)$ (i.e., the ideal generated by Q and the polynomials in ASC).

Theorem A.5. Let ASC be an irreducible asc chain with parameters $u_1, ..., u_q$. We can find an irreducible asc chain ASC' such that $PD(ASC) = PD(ASC')$ and the initials of the polynomials in ASC' are polynomials of the u.

Proof. It is a direct consequence of Lemma A.4. ∎

Theorem A.6. (Ritt–Wu's decomposition algorithm) For finite polynomial sets PS and DS, we can either detect the emptiness of $Zero(PS/DS)$ or find irreducible asc chains ASC_i, $i = 1, ..., l$, such that

$$Zero(PS/DS) = \cup_{i=1}^{l} Zero(PD(ASC_i)/DS)$$

The decomposition satisfies (a) there are no $i \neq j$ such that $PD(ASC_i) \subset PD(ASC_j)$; and (b) $prem(d, ASC_i) \neq 0$ for all $d \in DS$ and $i = 1, ..., l$.

Proof. See [Chou & Gao, 1990a,b]. ∎

References.

Abhyankar, S.S. and Bajaj, C. (1989), Computations with Algebraic Curves, *ISSAC-89*, LNCS No. 358, pp. 274–284, Springer-Verlag, 1989.

Abhyankar, S.S., Chandrsekar, S. and Chandru, V. (1990), Improper Intersection of Algebraic Curves, *ACM Tran. in Graphics*, 9(2), 147–159.

Arnon, D.S. and Sederberg, T.W. (1984), Implicit Equation for a Parametric Surface by Gröbner Bases, *Proc. 1984 MACSYMA User's Conference* (V.E. Golden ed.), General Electric, Schenectady, New York, 431–436.

Artin, M. and Mumford, D. (1972), Some Elementary Examples of Unirational Varieties Which Are Non-rational, *Proc. London Math. Soc.*, (3) 25, pp. 75-95.

Bajaj, C. (1990a) Rational Hypersurface Display, *Proc. of 1990 Symposium on Interactive 3D Graphics*, Snowbird, Utah.

Buchberger, B. (1985), Gröbner bases: an algorithmic method in polynomial ideal theory, *Recent Trends in Multidimensional Systems Theory* (ed. N.K. Bose), D.Reidel Publ. Comp., 1985.

Buchberger, B. (1987), Applications of Gröbner Bases in Non-linear Computational Geometry, L.N.C.S. No 296, R.JanBen (Ed.), pp. 52–80, Springer-Verlag.

Castelnuvo, (1894), Sulla Rationalita della Involuzioni Pinae, *Math. Ann.*, 44, pp. 125–155.

Chionh, E.W., (1990), *Base Points, Resultants, and the Implicit Representation of Rational Surfaces*, PhD Thesis, The Univ. of Waterloo.

Chou, S.C. (1988), *Mechanical Geometry Theorem Proving*, D.Reidel Publishing Company, 1988.

Chou, S.C. and Gao, X.S. (1990a), Ritt-Wu's Decomposition Algorithm and Geometry Theorem Proving, *10th International Conference on Automated Deduction*, M.E. Stickel (Ed.) pp 207–220, Lect. Notes in Comp. Sci., No. 449, Springer-Verlag.

Chou, S.C. and Gao, X.S. (1990b), Techniques for Ritt-Wu's Decomposition Algorithm, in*Proc. of IWMM'92*, International Academia Publishers, 1992.

Chuang, J.H., and Hoffman, C.M. (1989), On Local Implicit Approximation and Its Applications, *ACM Tran. in Graphics*, 8(4), pp. 298–324.

Faux, I.D. and Pratt, M.J. (1979), *Computational Geometry for Design and Manufacture*, Ellis Horwood, Chichester.

Farouki R.T. and Neff C.A., Automatic Parsing of degenerate Quadric-Surface Intersections, *ACM Trac. on Graphics, vol8, No. 3, 1989.*

Gallo, C. and Mishra B. (1990), Efficient Algorithms and Bounds for Wu-Ritt Characteristic Sets, TR, Courant Institute of Math.

Gao, X.S. and Chou, S.C. (1990a), On the Parameterization of Algebraic Curves, *Applicable Algebra in Elementary Communication and Computing*, 3, 27–38, Springer-Verlag, 1992.

Gao, X.S. and Chou, S.C. (1990b), On the Normal Parameterization of Curves and Surfaces, *The International Journal of Computational Geometry & Applications*, vol. 1, p.125-136, 1991, World Science Press.

Gao, X.S. and Chou, S.C. (1990c), Implicitization of Rational Parametric Equations, to appear in *Journal of Symbolic Computation.*

Gao, X.S. and Chou, S.C. (1990d), Computations with Parametric Equations, to appear at *Proc. of ISSAC'91*, ACM, New York.

Gröbner, W.(1970), *Algebraic Geometrie I, II,* Bibliographisches Institut, Mannheim.

Hartshorne, R.(1977), *Algebraic Geometry*, Springer-verlag.

Heintz, J. (1983), Definability and Fast Quantifier Elimination in Algebraically Closed Fields, Theoretic Computing Science, 24, 239–278.

Hodge, W.V.D. and Pedoe, D (1952), *Methods of Algebraic Geometry*, vol I and II, Cambridge.

Hoffmann, C.M. (1989), *Geometric and Solid Modeling: an Introduction*, Morgan Kaufmann Publishers Inc, 1989.

Hoffmann, C.M. (1990), Algebraic and Numeric Techniques for Offsets and Blends, in *Computations of Curves and Surfaces*, (eds. W. Dahman), p.499-529, Kluwer Academic Publishers.

Kalkbrener, M.(1990), Implicitization by Using Gröbner Bases, Technical Report RISC Series 90-27, Unin. of Linz.

Li, Z.M. (1987), On the Triangulation for Any Polynomial Set, MM-preprints, No. 2, p.48-54, Institute of systems science, Academia Sinica.

Li, Z.M. (1989), Automatic Implicitization of Parametric Objects, *MM Research Preprints*, No4, Ins. of Systems Science, Academia Sinica.

Manocha, D. and Canny J. F. (1990), Implicitizing Rational Parametric Surfaces, UCB/CSD, September.

Ollivier, F. (1989), Inversibility of Rational Mappings and Structural identifiability in Automatics, *Proc. of ISSAC-89*, p43-54, ACM Press.

Pratt, M.J. (1986), Parametric Curves and Surfaces Used in Computer Aided Design, in *The Mathematics of Surfaces* (ed. by J.A. Gregory), pp19–46, Clarendon Press, Oxford.

Sederberg, T.W. (1986), Improperly Parametrized Rational Curves, *Computer Aided Geometric Design*, vol. 3, pp. 67-75, 1986.

Sederberg, T.W., Anderson, D.C. and Goldman, R.N. (1984), Implicit Representation of Parametric Curves and Surfaces, *Computer Vision, Graph, Image Proc.*, vol28 pp 72–84.

Sederberg, T.W., Anderson, D.C. and Goldman, R.N. (1985), Implicitization, Inversion, and Intersection of Planar Rational Cubic Curves, Vision, Gragh, Image Proc. vol31 pp 89–102.

Sederberg, T.W. and Kakimoto, M. (1990), Approximation Rational Curves Using Polynomial Curves, ECGL 90-03, June 1990.

Segre, B. (1951), Sull Esistenza, Sia Nel Campo Rationale chenel Campo Reale, *Rend. Accad. Naz. Lincei* (8) 10, pp. 564–570.

Shannon, S. and Sweedler, M. (1988), Using Gröbner Bases to Determine Algebraic Membership, *J. Symbolic Computation*, 6, p.267-273.

Shoenfield, J. R. (1967), *Mathematical Logic*, Addison-Wesley.

Tarski, A, (1951), *A Decision Method for Elementary Algebra and Geometry*, Univ. of California Press, Berkeley, Calif., 1951.

Walker, R. (1950), *Algebraic Curves*, Princeton Univ. Press.

Wu, W.T. (1984), Basic Principles of Mechanical Theorem Proving in Elementary Geometries, *J. Sys. Sci. & Math. Scis.*, 4(1984), 207 –235, Re-published in *J. Automated Reasoning*, 1986.

Wu, W.T. (1986), On zeros of algebraic equations — an application of Ritt principle, *Kexue Tongbao*, 31(1986), 1–5.

Wu, W.T. (1989), On a Projection Theorem of Quasi-Varieties in Elimination Theory, *MM Research Preprints*, No. 4, Ins. of Systems Science, Academia Sinica.

Computer Mathematics
Proc. of the Special Program at
Nankai Institute of Mathematics
January 1991 - June 1991

Computer versus paper and pencil

Maurice Mignotte

Université Louis Pasteur
Mathématique
67084 Strasbourg, FRANCE

1. Introduction

This paper corresponds to a lecture given in the NANKAI Institute in Tianjin, April 1991. The results given here were obtained in 1985, but never published.

The main reason to present them is the fact that we use only very elementary arguments. Because of their simplicity, these arguments can be used in other problems of quantifier elimination. Or maybe, they can be combined with the tools offered by systems of computer algebra.

Moreover, this work had some influence on the study on this field . This can be seen in the references [4] and [5].

Besides, the solutions we get by these easy arguments are simpler than those which were produced by cylindrical decomposition (which splits the initial problem into many subcases).

We have followed the initial manuscript without any changes.

2. First example

The following problem, say **P1**, has been considered in [1]: find conditions on the real parameters a, b, c and d such that the ellipse

$$\frac{(x-c)^2}{a^2} + \frac{(y-d)^2}{b^2} = 1 \quad (with \quad a > 0, \quad b > 0)$$

is contained in the unit disk

$$x^2 + y^2 \leq 1.$$

In [1], only the special case $d = 0$ could be solved using a quantifier elimination algorithm. Here is a non-assisted solution in this special case.

Put

$$g(x,y) = \frac{(x-c)^2}{a^2} + \frac{(y-d)^2}{b^2} - 1, \quad f(x,y) = x^2 + y^2 - 1.$$

The problem **P1** is equivalent to the condition

$$(1) \qquad g(x,y) = 0 \Longrightarrow f(x,y) \le 1.$$

Since Lagrange, we know that if the point (x, y) corresponds to an extremum of a function f on some curve g=0 then the gradients of f and g at this point are collinear; here it implies

$$\frac{x-c}{a^2} = \lambda x \quad and \quad \frac{y-d}{b^2} = \lambda y,$$

for a certain real number λ.

Notice the relations

$$(2) \qquad (1 - \lambda a^2)x = c, \quad (1 - \lambda b^2)y = d.$$

Suppose first that

$$(i) \qquad c = d = 0$$

Then **P1** reduces to

$$\frac{x^2}{a^2} + \frac{y^2}{b^2} = 1 \Longrightarrow x^2 + y^2 \le 1,$$

and the previous relations show that, except in the trivial case $a^2 = b^2$, we only have to consider the case $xy = 0$. We conclude that, in this case, **P1** is equivalent to the condition

$$max\{a^2, b^2\} \le 1,$$

which is also obviously the condition to be satisfied when $a^2 = b^2$.

Now, suppose that

$$(ii) \qquad c \ne 0 \text{ and } d = 0.$$

By symmetry, we may assume $c > 0$. Relations (2) show that $\lambda = b^{-2}$ or $y = 0$.

When $y = 0$, **P1** is equivalent to

$$(x-c)^2 = a^2 \Longrightarrow x^2 \le 1,$$

thus we find the condition

$$a + c \le 1.$$

The case $\lambda = b^{-2}$ leads to the relation

$$x = \frac{cb^2}{b^2 - a^2},$$

so that – if y exists – we have

$$y^2 = b^2\left(1 - \frac{(x-c)^2}{a^2}\right) = b^2\left(1 - \frac{a^2c^2}{(b^2-a^2)^2}\right),$$

and the condition $x^2 + y^2 \leq 1$ becomes

$$c^2b^4 + b^2((b^2-a^2)^2 - c^2a^2) \leq (b^2-a^2)^2,$$

or

$$(b^2-a^2)b^2(c^2+b^2-a^2) \leq (b^2-a^2)^2.$$

To summarize, we have proved the following result.

THEOREM 1. – *The ellipse*

$$\frac{(x-c)^2}{a^2} + \frac{(y-d)^2}{b^2} = 1 \quad (a > 0, b > 0)$$

is contained in the unit disk $x^2 + y^2 \leq 1$ if, and only if, one of the following conditions holds

- $(c = 0)$ and $(a^2 \leq 1)$ and $(b^2 \leq 1)$,
- $(c \neq 0)$ and $(a + |c| \leq 1)$ and $((b^2-a^2)^2 < a^2c^2)$,
- $(c \neq 0)$ and $(a + |c| \leq 1)((b^2-a^2)^2 \geq a^2c^2)$ and $(b^2(b^2-a^2)(b^2+c^2-a^2) \leq (b^2-a^2)^2)$.

Remarks

1) In the general case $cd \neq 0$, we have proved that **P1** is equivalent to the condition

$$\frac{a^2c^2\lambda^2}{(1-\lambda a^2)^2} + \frac{b^2d^2\lambda^2}{(1-\lambda b^2)^2} = 1 \Longrightarrow \frac{c^2\lambda^2}{(1-\lambda a^2)^2} + \frac{d^2\lambda^2}{(1-\lambda b^2)^2} \leq 1,$$

which contains only one variable to eliminate.

2) As indicated in [1], a computer-assisted proof of the general case is given in [3]; but this paper is not available to me.

3. Second example

The problem considered in [2] is: find conditions on p,q and r such that

$$\forall x,\quad x^4 + px^2 + qx + r \geq 0, \tag{3}$$

and a solution was obtained which needed 25 minutes of C.P.U. time.

Firstly we consider the case

$$r = 0. \tag{i}$$

Then (3) is equivalent to the two conditions

$$\begin{cases} x \leq 0 \Longrightarrow x^3 + px + q \leq 0, \\ x \geq 0 \Longrightarrow x^3 + px + q \geq 0, \end{cases}$$

which appear to be equivalent to

$$q = 0 \quad \text{and} \quad p \geq 0.$$

Suppose now that

$$r \neq 0, \tag{ii}$$

then $r > 0$ (consider $x = 0$). Put $\widetilde{p} = p/r$, $\widetilde{q} = q/r$ and $\widetilde{r} = 1/r$. Then (3) is equivalent to

$$\forall y, Q(y) > 0, \qquad where \qquad Q(y) = y^4 + \widetilde{q}y^3 + \widetilde{p}y^2 + \widetilde{r}. \tag{4}$$

Since $Q(y)$ tends to $+\infty$ with $|y|$, we have to prove that $\min\{Q(y)\} \geq 0$, *i.e.*

$$4y^3 + 3\widetilde{q}y^2 + 2\widetilde{p}y = 0 \Longrightarrow Q(y) \geq 0. \tag{5}$$

Changing the sign of y if necessary, we may suppose $\widetilde{q} \geq 0$. By Euclidean division we obtain

$$Q(y) \equiv Ay + B \quad (mod \quad 4y^2 + 3\widetilde{q}y^2 + 2\widetilde{p}y)$$

where

$$A = \frac{9\widetilde{q}^3}{64} - \frac{\widetilde{p}\widetilde{q}}{2} \text{ and } B = \widetilde{r} + \frac{3\widetilde{p}\widetilde{q}^2}{32} - \frac{\widetilde{p}^2}{4}$$

This shows that (5) is equivalent to

$$(\widetilde{r} > 0) \text{ and } (4y^2 + 3\widetilde{q}y^2 + 2\widetilde{p}y = 0 \Longrightarrow Ay + B \geq 0). \tag{6}$$

Put $\widetilde{\Delta} = 9\widetilde{q}^2 - 32\widetilde{p}$, then (6) splits into two case:

$$(\widetilde{r} > 0) \text{ and } (\Delta < 0), \tag{6'}$$

and

$$(6") \qquad (\tilde{r} > 0) \text{ and } (\tilde{\Delta} \geq 0) \text{ and } \left(y = \frac{-3\tilde{q} - \sqrt{\tilde{\Delta}}}{8} \Longrightarrow Ax + B \geq 0\right).$$

The relation $A = \frac{1}{64}\tilde{q}\tilde{\Delta}$, with $\tilde{q} \geq 0$, shows that (6") is equivalent to

$$(\tilde{r} > 0) \text{ and } (\tilde{\Delta} \geq 0) \text{ and } \left(\tilde{q}\tilde{\Delta}\left(\frac{-3\tilde{q} - \sqrt{\tilde{\Delta}}}{8}\right) + 64B \geq 0\right),$$

which simplifies to

$$(\tilde{r} > 0) \text{ and } (\tilde{\Delta} \geq 0) \text{ and } \left(18\tilde{q}^2\tilde{p} + 64\tilde{r} - \frac{\tilde{q}\tilde{\Delta}^{3/2}}{8} \geq 16\tilde{p}^2 + \frac{27}{8}\tilde{q}^4\right).$$

Thus, we have proved the following proposition.

THEOREM 2.–*The polynomial*

$$x^4 + px^2 + qx + r \quad \textit{where} \quad q \geq 0,$$

is never negative if, and only if , one of the following conditions holds

- $(q = r = 0)$ and $(p \geq 0)$,
- $(r > 0)$ and $(9q^2 < 32pr)$,
- $(r > 0)$ and $(\Delta := 9q^2 - 32pr \geq 0)$
  and $\quad (18q^2pr + 64r^3 - \frac{1}{8}q\Delta^{\frac{3}{2}} \geq 16p^2r^2 + \frac{27}{8}q^4)$.

Remarks

3) Trivial modifications of the proof of theorem 2 lead to the solution of the same problem for the polynomials $x^4 + lx^3 + px^2 + qx + r$ or $x^{2n} + px^2 + qx + r$, where n is an integer ≥ 2.

4) In the first problem, the ellipse

$$\frac{(x-c)^2}{a^2} + \frac{(y-d)^2}{b^2} = 1$$

can be parametrized by

$$x = c + \frac{2at}{1+t^2} \text{ and } y = d + \frac{(1-t^2)b}{1+t^2},$$

so that the ellipse is contained in the unit disk if , and only if , we have

$$\forall t, \left(c + \frac{2at}{1+t^2}\right)^2 + \left(d + \frac{(1-t^2)b}{1+t^2}\right)^2 \leq 1,$$

or, equivalently, if

$$\forall t, (1+t^2)^2 - ((1+t^2)^2c + 2at)^2 - ((1+t^2)^2d + (1-t^2)b)^2 \geq 0.$$

By remark 2, it is possible—in principle— to determine conditions on a, b, c and d such that this assertion is true.

References

[1] D.S. ARNON, S.F. SMYTH. - Towards a mechanical solution of the ellipse problem I, Computer Algebra, EUROCAL'83, London, ed. J.A. van Hulzen, *Lecture Notes in Computer Science*, 162, Springer, Berlin,1983

[2] D.S. ARNON, S.F. SMYTH. - On the mechanical Quantifier Elimination for Elementary Algebra and Geometry; EUROCALINZ 85, Linz, April 1985.

[3] M. LAUER. -A solution to Kahan's ellipse problem, *SIGSAM Bull. of the Assoc. Comp. Mach.*, 11, 1977, pp. 16-20.

[4] D.S. ARNON, M. MIGNOTTE. -On Mechanical Quantifier Elimination for Elementary Algebra and Geometry; *J. of Symbolic Comp.*, 5, 1988, pp. 237-259.

[5] D. LAZARD. -Quantifier Elimination: Optimal Solution for two Classical Examples; *J. of Symbolic Comp.*, 5, 1988, pp. 261-266.

Computer Mathematics
Proc. of the Special Program at
Nankai Institute of Mathematics
January 1991 - June 1991

The Finite Basis of an Irreducible Ascending Set

Shi, He

Institute of Systems Science, Academia Sinica
Beijing 100080, China

Let K be a field of characteristic 0 and $u_1, u_2, \cdots, u_d, x_1, x_2, \cdots, x_n$ be a set of indeterminates. We consider an irreducible ascending set

$$(IRR) \quad A_n, A_{n-1}, \cdots, A_1,$$

in which every A_i is a polynomial in the ring $K[u_1, u_2, \cdots, u_d, x_1, x_2, \cdots, x_i]$ and has the following form

$$A_i = I_i * x_i^{d_i} + lower\ degree\ terms\ in\ x_i.$$

The notions of the irreducible ascending set and the generic point determined by it were introduced in [WU1,WU2]. In [WU3], Professor Wu Wen-tsun discussed the notions of an irreducible ascending set, an affine algebraic variety, the generic point and the Chow basis and pointed out that they are equivalent. As in [WU3], the ideal of polynomials having ξ as zeros will be denoted by Ideal$[IRR]$,where ξ is the generic point of the irreducible ascending set $IRR : A_n, A_{n-1}, \cdots, A_1$.

By the Hilbert Finite-Basis Theorem, the Ideal[IRR] has a finite basis necessarily. The Chow form, which is called the Cayley form in [H-P], furnished a method to determine explicitly a finite basis, called Chow basis of Ideal[IRR](see [WU3]). However, it is not a simple computational procedure to obtain the Chow basis via the Chow form.

The pseudo-resultant formula of a given polynomial with an ascending set is introduced in [SH]. In this paper an algorithm to determine a finite basis of an irreducible ascending set is provided by using the pseudo-resultant formula. In Section 1, we discuss the pseudo-resultant formula of an irreducible ascending set with a polynomial. The finite basis of Ideal [IRR] is defined and an algorithm to compute it via the pseudo-resultant formula is provided in Section 2. And in Section 3 an illustrated example is discussed.

1. *The Pseudo-resultant Formula*

In this section, we introduce the pseudo-resultant formula of a polynomial with an irreducible ascending set.

Let F_1, F_2 be two polynomials in the ring $K[u, x_1, x_2, \cdots, x_i]$ with the same main variable x_i. The degrees of F_1, F_2 in x_i are $deg_i(F_1) = m_1, deg_i(F_2) = m_2$. The pseudo-resultant of F_1 and F_2 can be obtained by the division algorithm of two polynomials as follows:

$$
\begin{aligned}
I_2 * F_1 &= b_1 * F_2 + M_1,\\
c_1 * F_2 &= b_2 * M_1 + M_2,\\
\cdots \quad &\cdots,\\
c_k * M_{k_1} &= b_{k+1} * M_k + RS,
\end{aligned}
$$

in which I_2 is the initial of F_2; $c_1, c_2, \cdots, c_k$ are the initials of $M_1, M_2, \cdots, M_k$, and RS is the pseudo-resultant of F_1 and F_2. Then, after eliminating $M_k, M_{k-1}, \cdots, M_2, M_1$ successively we obtain

$$a_1 * F_1 + a_2 * F_2 = RS, \tag{1.1}$$

in which a_1, a_2 are polynomials in the ring $K[u, x_1, x_2, \cdots, x_i]$ with

$$deg_i(a_1) < deg_i(F_2), deg_i(a_2) < deg_i(F_1), \tag{1.2}$$

and RS is a polynomial in the ring $K[u, x_1, x_2, \cdots, x_{i-1}]$.

Let $AS_n : A_n, A_{n-1}, \cdots, A_1$ be an irreducible ascending set. For a given polynomial G, we consider the pseudo-resultant of G with AS_n. First, we compute the pseudo-resultant of G with A_n by the division algorithm

$$
\begin{aligned}
I_n * G &= b_1 * A_n + M_1,\\
c_1 * A_n &= b_2 * M_1 + m_2,\\
\cdots \quad &\cdots,\\
c_k * M_{k-1} &= b_{k-1} = b_{k+1} * M_k + RS.
\end{aligned}
$$

For an ascending set AS_n, the case of $A_{n-1} = 0, \cdots, A_1 = 0$ is usually considered. For example, the polynomials $A_n, \cdots, A_1$ are equal to zero if the indeterminates $u_1, \cdots, u_d, x_1, x_2, \cdots, x_n$ are substituted by the generic point ξ of the irreducible ascending set. In order to guarantee the efficiency of the division algorithm, it is necessary to avoid the annihilating of the initials I_n, $c_1; c_2, \cdots, c_k$ when $A_{n-1} = 0, \cdots, A_1 = 0$. In the case of irreducible ascending set, it is sufficient to consider that the polynomials $M_1, M_2, \cdots, M_j$ are reduced with respect to the ascending set $AS_{n-1} : A_{n-1}, A_{n-2}, \cdots, A_1$. Therefore, we should consider the reduction of the remainders $M_1, M_2, \cdots, M_j$. The procedure of the division algorithm becomes

$$I * G = d_1 * A_n + N_1 + \sum_{i=1}^{n-1}(P_{1i} * A_i),$$

$$e_1 * A_n = d_2 * N_1 + N_2 + \sum_{i=1}^{n-1}(p_{2i} * A_i),$$

$$\cdots \quad \cdots \quad \cdots,$$

$$e_j * N_{j-1} = d_{j+1} * N_j + RS_{n-1},$$

in which I is a product of some power of initials of $A_n, A_{n-1}, \cdots, A_1$; $e_k, k = 1, 2, \cdots, j$, are some products of the power of initials of $N_k, A_n, A_{n-1}, \cdots, A_1$; and $N_1, N_2, \cdots, N_j$ are reduced with respect to the ascending set AS_{n-1}. Now, by eliminating $N_j, N_{j-1}, \cdots, N_1$ successively we obtain

$$J_n * G + B_n * A_n = RS_{n-1} + \sum_{i}^{n-1}(Q_i(n) * A_i), \tag{1.3}$$

in which the degrees of J_n , B_n satisfy

$$deg_n(J_n) < deg_n(A_n), deg_n(B_n) < deg_n(G).$$

Similarly to this procedure, we can get the pseudo-resultant of polynomials RS_{n-1} and A_{n-1}

$$J_{n-1} * RS_{n-1} + B_{n-1} * A_{n-1} = RS_{n-2} + \sum_{i=1}^{n-2}(Q_i(n-1) * A_i).$$

And we obtain successively the pseudo-resultants of RS_j with A_j as follows:

$$J_i * RS_j + B_j * A_j = RS_{j-1} + \sum_{i=1}^{j-1}(Q_i(j) * A_i),$$
$$j = n-1, n-2, \cdots, 1,$$

where the procedures of the division algorithm are

$$\begin{aligned} I(j) * RS_j &= d_{1j} * A_j + N_{1j} + \sum_{i=1}^{j-1}(p_{1j_i} * A_i), \\ e_{1j} * A_j &= d_{2j} * N_{1j} + N_{2j} + \sum_{i=1}^{j-1}(p_{2j_i} * A_i), \\ &\cdots \quad \cdots \quad \cdots, \\ e_{kj} * N_{k,j-1} &= d_{k,j+1} * N_{kj} + RS_{j-1}, \\ j &= n-1, n-2, \cdots, 1. \end{aligned} \tag{1.5}$$

In formula (1.5), the remainders $N_{1j}, N_{2j}, \cdots, N_{kj}$ are reduced with respect to the irreducible ascending set $AS_{j-1} : A_{j-1}, \cdots, A_1$; and $I(j), e_{1j}, \cdots, e_{kj}$ are some products of the power of initials of $A_j, A_{j-1}, \cdots, A_1, N_{1j}, N_{2j}, \cdots, N_{kj}$. Now by eliminating successively the pseudo-resultants $RS_1, RS_2, \cdots, RS_{n-1}$ we obtain

$$J * G + \sum_{j=1}^{n} (P_j * A_j) = RS_0, \tag{1.6}$$

in which $J = J_1 * J_2 * \cdots * J_n$ and the degrees of $J_i, i = 1, 2, \cdots, n$, satisfy

$$deg_i(J_i) < deg_i(A_i). \tag{1.7}$$

The polynomial RS_0 is determined by both G and the irreducible ascending set AS_n. It is called the *pseudo-resultant* of G with AS_n. The formula (1.6) is called the *pseudo-resultant formula* of polynomial G with the irreducible ascending set AS_n.

Remark 1.1. The resultant of two polynomials F, G is well-defined (see [H-P]). By definition, it is a determinant of the coefficients of F and G. In this paper, the pseudo-resultant of two polynomials is obtained by the division algorithm. The difference between the resultant and the pseudo-resultant of two polynomials is a certain factor. It was studied by Collins (see [CO]).

Let $\xi = (u_1, u_2, \cdots, u_d, \xi_1, \xi_2, \cdots, \xi_n)$ be a generic point of the irreducible ascending set AS_n. We have the conclusion

Theorem 1.1. For a given polynomial G and an irreducible ascending set AS_n : $A_n, A_{n-1}, \cdots, A_1$ as above , the pseudo-resultant formula (1.6) is efficient when the indeterminates $u_1, u_2, \cdots, u_d, x_1, x_2, \cdots, x_i$ are replaced by $u_1, u_2, \cdots, u_d, \xi_1, \xi_2, \cdots, \xi_n$.

In fact, since the remainders $N_{1j}, N_{2j}, \cdots, N_{kj}$ in formula (1.5) are reduced with respect to ascending set $AS_{j-1} : A_{j-1}, \cdots, A_1$, the initials $A_j, A_{j-1}, \cdots, A_1, N_{1k}, N_{2j}, \cdots, N_{kj}$, are not equal to zero when the indeterminates $u_1, u_2, \cdots, u_d, x_1, x_2, \cdots, x_i$ are replaced by $u_1, u_2, \cdots, u_d, \xi_1, \xi_2, \cdots, \xi_n$. Therefore, the factors $I(j), e_{1j}, \cdots, e_{kj}$ are not equal to zero neither.

Theorem 1.2. For a polynomial $G(u, x)$ in the ring $K[u, x]$, the generic point ξ of an irreducible ascending set AS_n is a zero of G if and only if the pseudo-resultant RS_0 of G with AS_n is 0.

Proof. If $G(\xi) = 0$, then from the pseudo-resultant formula (1.6) we know that the pseudo-resultant $RS_0(\xi)$ equals 0. However, RS_0 is reduced with respect to the irreducible ascending set AS_n. Thus, the pseudo-resultant RS_0 must be zero.

Otherwise, if the pseudo-resultant RS_0 is 0, then by the pseudo-resultant formula (1.6) we know that $J(\xi) * G(\xi)$ equals zero. But the relations (1.7) mean that the factors $J_i, i = 1, 2, \cdots, j-1$, are reduced with respect to the irreducible ascending set AS_j .Therefore $J_i(\xi), i = 1, 2, \cdots, j-1$, are not equal to zero. Hence $G(\xi)$ equals zero. The theorem is proved.

2. *The Finite Basis*

In this section we introduce a finite basis of the Ideal [IRR] by using the pseudo-resultant formula (1.6) provided in Section 1.

Let $AS_n : A_n, A_{n-1}, \cdots, A_1$ be an irreducible ascending set and

$$\xi = (u_1, u_2, \cdots, u_d, \xi_1, \xi_2, \cdots, \xi_n)$$

be the generic point of AS_n. We set

$$v_0 = -\sum_{i=1}^{n}(v_i * \xi_i), \tag{2.1}$$

where v_i, $i = 1, 2, \cdots, n$, are parameters. We consider the linear function $H(x, \xi, v)$ of $x_1, \cdots, x_n$ defined as

$$\begin{aligned} H(x, \xi, v) &= \sum_{i=1}^{n}(v_i * x_i) + v_0 \\ &= \sum_{i-1}^{n}(v_i * x_i - v_i * \xi_i). \end{aligned} \tag{2.2}$$

It is trivial that $H(\xi, \xi, v) = 0$. Certainly, this linear function $H(x, \xi, v)$ can be considered as a polynomial in the ring $K[u, x_1, x_2, \cdots, x_n]$. By using the division algorithm provided in Section 1 we can obtain the pseudo-resultant of $H(x, \xi, v)$ and AS_n as

$$J * H(x, \xi, v) + \sum_{i=1}^{n}(P_i * A_i) = RS_0(u, \xi, v), \tag{2.3}$$

where the pseudo-resultant $RS_0(u, \xi, v)$ is a polynomial of the parameters $v_1, \cdots, v_n$, and $u_1, u_2, \cdots, u_d, \xi_1, \xi_2, \xi_n$. It can be written as

$$RS_0(u, \xi, v) = \sum_{i_1, i_2, \cdots, i_n} B_{i_1 i_2 \cdots i_n}(u, \xi) * v_1^{i_1} v_2^{i_2} \cdots v_n^{i_n}, \tag{2.4}$$

where $B_{i_1 i_2 \cdots i_n}$ are polynomials of $u_1, u_2, \cdots, u_d, \xi_1, \xi_2, \cdots, \xi_n$.

From Theorem 1.2 we know that the pseudo-resultant $RS_0(u, \xi, v)$ is zero when the indeterminates $x_1, \cdots, x_n$ are replaced by $\xi_1, \xi_2, \cdots, \xi_n$ in formula (2.3). However, the pseudo-resultant $RS_0(u, \xi, v)$ is independent of $x_1, \cdots, x_n$. Thus we have

Lemma 2.1. Let the linear function $H(x, \xi, v)$ be given in (2.2) and $RS_0(u, \xi, v)$ be the pseudo-resultant of $H(x, \xi, v)$ with the irreducible ascending set AS_n. Then the generic point ξ of AS_n is a zero of the polynomials

$$B_{i_1 i_2 \cdots i_n}(u, x)$$

which are given in formula (2.4).

Proof. From the pseudo-resultant formula we know that the pseudo-resultant $RS_0(u,\xi,v)$ is zero. Now by the arbitrariness of the parameters $v_1, v_2, \cdots, v_n$, we get

$$B_{i_1 i_2 \cdots i_n}(u,\xi) = 0.$$

The lemma is proved.

The extension field by adjoining the indeterminates $u_1, u_2, \cdots, u_d$ to the field will be denoted by $K(u)$. The irreducible algebraic variety determined by the generic point ξ of the irreducible ascending set AS_n will be denoted by

$$Var[\xi] \subseteq Var(B_i). \tag{2.5}$$

We denote the generic point of variety $Var(B_i)$ by η if $Var(B_i)$ is also an irreducible one. By definition, it means $B_{i_1 i_2 \cdots i_n}(u,\eta) = 0$. From pseudo-resultant formula (2.3) we know that $A_i(u,\eta) = 0, i = 1, 2, \cdots, n$. That is, η is a specialization of generic point ξ. Thus, two varieties are equivalent:

$$Var[\xi] = Var(B_i).$$

Definition 2.1. The set of polynomials $B_{i_1 i_2 \cdots i_n}(u,x)$ is a finite basis of the Ideal $[IRR]$ if the variety $Var(B_i)$ over the field $K(u)$ defined by them is an irreducible algebraic variety.

Summing up the discussions above, we have

Theorem 2.2. For a given irreducible ascending set $AS_n : A_n, \cdots, A_1$, the pseudo-resultant $RS_0(u,\xi,v)$ of AS_n with the linear functions $H(x,\xi,v)$ given by formula (2.2) should be zero. When the algebraic variety $Var(B_i)$ over the extension field $K(u)$ is an irreducible variety, the set of polynomials given in the expression (2.4) of the pseudo-resultant $RS_0(u,\xi,v)$

$$B_{i_1 i_2 \cdots i_n}(u,x) \tag{2.6}$$

provides a finite basis of the Ideal[IRR].

Corollary 2.3. If the set of polynomials $B_{i_1 i_2 \cdots i_n}(u,x)$ in (2.6) forms an irreducible ascending set, then it is a finite basis of Ideal[IRR].

Remark 2.1. From the definition of the resultant of two polynomials, the resultant formula of a polynomial G with an irreducible ascending set AS_n can be defined (see [CH]). Similarly to this paper, the finite basis of an irreducible ascending set can be discussed by using the resultant formula.

3. *The Desargues Configuration*

In [WU3] , as an example to illustrate the general decomposition theory established by Wu, the Desargues Theorem is studied. There are two irreducible ascending sets introduced

$$ASC_1 : A_{13}, A_{12}, A_{11},$$
$$ASC_2 : A_{23}, A_{22}, A_{21}.$$

The polynomials in ASC_1 are

$$\begin{aligned} A_{13} &= u_4 * x_3 - (u_5 - u_3) * x_2 - u_4 * x_1, \\ A_{12} &= (u_5u_1 + u_4u_3 - u_3u_1) * x_2 - (u_4^2 - u_4u_1) * x_1 - u_5u_4u_2, \\ A_{11} &= u_1 * x_1 - u_3u_2 \end{aligned} \tag{3.1}$$

and the polynomials in ASC_2 are

$$\begin{aligned} A_{23} &= u_4 * x_3 - u_4 * x_2 - u_5 * x_1 + u_5u_3, \\ A_{22} &= u_1 * x_2 - u_3u_2, \\ A_{21} &= u_1 * x_1 + u_4u_3 - u_3u_1. \end{aligned} \tag{3.2}$$

The Chow basis of Ideal $[ASC_1]$ is obtained as

$$\begin{aligned} CB_{11} &= u_1 * x_1 - u_3u_2, \\ CB_{12} &= u_1 * x_2 - u_4u_2, \\ CB_{13} &= u_1 * x_3 - u_5u_2. \end{aligned} \tag{3.3}$$

and the Chow basis of Ideal $[ASC_2]$ is obtained as

$$\begin{aligned} CB_{21} &= -u_1^2 * x_2 + u_1u_2 * x_1 + u_4u_3u_2, \\ CB_{22} &= u_1^2 * x_3 - u_1u_2 * x_1 + u_5u_3u_1 - u_4u_3u_2, \\ CB_{23} &= -u_1 * x_3 + u_1 * x_2 - u_5u_3, \\ CB_{24} &= u_1 * x_3 + u_5u_3 - u_3u_2, \\ CB_{25} &= u_1 * x_1 + u_4u_3 - u_3u_1, \\ CB_{26} &= u_1 * x_2 - u_3u_2. \end{aligned} \tag{3.4}$$

Remark 3.1. It is easy to check that there are some relations between the Chow basis (3.4). In fact, the following equations are valid:

$$\begin{aligned} CB_{21} &= CB_{25} * u_2 - CB_{26} * u_1, \\ CB_{22} &= CB_{24} * u_1 - CB_{25} * u_2, \\ CB_{23} &= -CB_{24} + CB_{26}. \end{aligned}$$

That is, the polynomials $CB_{21}, CB_{22}, CB_{23}$ are linear combinations of the polynomials $CB_{24}, CB_{25}, CB_{26}$. It means that the Chow basis includes usually some more polynomials.

The computation of the Chow basis via the Chow form is not a simple procedure. Therefore, a simple algorithm to determine the bases of Ideal $[IRR]$ is needed. Now, as an example, we compute the finite basis, which is defined in Section 2, of Ideal $[ASC_1]$ and Ideal $[ASC_2]$ by using the pseudo-resultant formula.

For the case of Ideal $[ASC_1]$, we set

$$H_1(x,\xi,v) = v_1 * x_1 + v_2 * x_2 + v_3 * x_3 + v_0, \tag{3.5}$$

where v_1, v_2, v_3 are parameters and

$$v_0 = -v_1 * \xi_1 - v_2 * \xi_2 - v_3 * \xi_3. \tag{3.6}$$

The pseudo-resultant of linear function $H_1(x,\xi,v)$ with the irreducible ascending set ASC_1 can be obtained by the pseudo-resultant formula as follows:

$$\begin{aligned} RS_{01}(u,\xi,v) &= u_5u_2v_3 + u_4u_2v_2 + u_3u_2v_1 + u_1v_0, \\ &\quad (u_5u_2 - u_1\xi_3) * v_3 + (u_4u_2 - u_1\xi_2) * v_2 + (u_3u_2 - u_1\xi_1) * v_1. \end{aligned}$$

From the arbitrariness of the parameters v_3, v_2, v_1 we get

$$\begin{aligned} u_1 * \xi_1 - u_3u_2 &= 0, \\ u_1 * \xi_2 - u_4u_2 &= 0, \\ u_1 * \xi_3 - u_5u_2 &= 0. \end{aligned}$$

Thus, by definition, the finite basis of Ideal $[ASC_1]$ is

$$\begin{aligned} FB_{11} &= u_1 * x_1 - u_3u_2, \\ FB_{12} &= u_1 * x_2 - u_4u_2, \\ FB_{13} &= u_1 * x_3 - u_5u_2. \end{aligned} \tag{3.7}$$

For the case of Ideal $[ASC_2]$, we set again

$$H_2(x,\xi,v) = v_1 * x_1 + v_2 * x_2 + v_3 * x_3 + v_0, \tag{3.8}$$

where v_1, v_2, v_3 are parameters and

$$v_0 = -v_1 * \xi_1 - v_2 * \xi_2 - v_3 * \xi_3. \tag{3.9}$$

The pseudo-resultant of linear function $H_2(x,\xi,v)$ with the irreducible ascending set ASC_2 can be obtained by the pseudo-resultant formula as follows:

$$\begin{aligned} RS_{02}(u,\xi,v) &= (u_5u_3 - u_3u_2) * v_3 - u_3u_2 * v_2 \\ &\quad + (u_4u_3 - u_3u_1) * v_1 - u_1v_0 \\ &= (u_5u_3 - u_3u_2 + u_1\xi_3) * v_3 - (u_3u_2 - u_1\xi_2) * v_2 \\ &\quad + (u_4u_3 - u_3u_1 + u_1\xi_1) * v_1. \end{aligned}$$

From the arbitrariness of the parameters v_3, v_2, v_1 we get

$$u_1 * \xi_1 + u_4 u_3 - u_3 u_1 = 0,$$
$$u_1 * \xi_2 - u_3 u_2 = 0,$$
$$u_1 * \xi_3 + u_5 u_3 - u_3 u_2 = 0.$$

Thus, by definition, the finite basis of Ideal $[ASC_2]$ is

$$\begin{aligned} FB_{21} &= u_1 * x_1 + u_4 u_3 - u_3 u_1, \\ FB_{22} &= u_1 * x_2 - u_3 u_2, \\ FB_{23} &= u_1 * x_3 + u_5 u_3 - u_3 u_2. \end{aligned} \tag{3.10}$$

In this example, the finite basis of an irreducible ascending set is simpler than Chow basis. And the computation of the finite basis via the pseudo-resultant formula is much simpler than the computation of Chow basis via Chow form.

REFERENCES

[CH] S.C. Chou, *Mechanical Geometry Theorem Proving*. D.Reidel Publishing Company, Pordrecht, Netherlands, 1988.

[CO] G.E. Collins, *Subresultants and reduced polynomial sequences*, J. ACM, 14, (1967), 128-142.

[HP] W.V.D.Hodge and D.Pedoe, *Methods of Algebraic Geometry*, Cambridge, 1952.

[SH] Shi He, *On the pseudo-resultant formula for mechanical theorem proving*, MM-Preprints, 4, 1989.

[WU1] Wu Wen-tsun, *Basic Principles of Mechanical Theorem Proving in Geometries (Part on Elementary Geometries)*(in Chinese), Science Press, 1984.

[WU2] Wu Wen-tsun, *Basic principles of mechanical theorem-proving in elementary geometries*, J. Sys. Sci. and Math. Sci., 4(1984) 207-235. Republished in J. of Automated Reasoning, 2 (1986) 221-252.

[WU3] Wu Wen-tsun, *On the generic zero and Chow basis of an irreducible ascending set*, MM-Preprints, 4, 1989.

Computer Mathematics
Proc. of the Special Program at
Nankai Institute of Mathematics
January 1991 - June 1991

A Note on Wu Wen-Tsün's Non-degenerate Condition

Zhang Jingzhong, Yang Lu, and Hou Xiaorong
Institute of Mathematical Sciences, Academia Sinica
610015 Chengdu, Sichuan, P.R. China

Abstract

It is shown that Wu's non-degenerate condition which plays an important role in Wu's theory on algebraic systems and automated theorem proving can be replaced by a much weaker one, so the characteristic-set-based method is extensively valid except for very few trivial instances.

1 Introduction

Given a system of algebraic equations PS,

$$\begin{cases} F_1(u_1, u_2, \cdots, u_d, x_1, x_2, \cdots, x_s) = 0, \\ F_2(u_1, u_2, \cdots, u_d, x_1, x_2, \cdots, x_s) = 0, \\ \cdots\cdots \\ F_s(u_1, u_2, \cdots, u_d, x_1, x_2, \cdots, x_s) = 0, \end{cases} \tag{1}$$

where u_i and x_j are regarded as parameters and indeterminates for $i = 1, \cdots, d$ and $j = 1, \cdots, s$, respectively, by *Ritt-Wu's well-ordering principle*[1][4-6], we can find an ascending chain AS, that is, the characteristic set of PS,

$$\begin{cases} f_1(u_1, u_2, \cdots, u_d, x_1) = 0, \\ f_2(u_1, u_2, \cdots, u_d, x_1, x_2) = 0, \\ \cdots\cdots \\ f_s(u_1, u_2, \cdots, u_d, x_1, x_2, \cdots, x_s) = 0. \end{cases} \tag{2}$$

Denote by $\text{lcoeff}(f_j, x_j)$ or l_j the leading coefficient of f_j with respect to x_j, for $j = 1, \cdots, s$. The condition

$$l_1 l_2 \cdots l_s \neq 0 \tag{3}$$

is called *Wu's non-degenerate condition* which plays an important role in Wu's theory on algebraic system and automated theorem proving[1][6]. Any zero of PS must be a zero of AS; conversely, if a zero of AS satisfies Wu's non-degenerate condition, then it must be a zero of PS. On proving an equality-type theorem, if the *final pseudo remainder* of the conclusion polynomial with respect to hypothesis polynomials vanishes, then the conclusion holds under the non-degenerate condition. Relevant improvements and discusions[2][3] appeared recently. Chou and Gao[2] pointed out that, for a class of "constructive geometry theorems", one need only check the condition for half the leading coefficients of hypothesis polynomials.

It is shown in this paper that Wu's non-degenerate condition can be weakened greatly and Wu's characteristic-set-based method for automated theorem proving is extensively valid except for very few trivial instances.

Throughout this paper, by the field we mean a complex field C unless a further claim is made. A common zero of $f_1, f_2, \cdots, f_j$ is called a $j-$zero, and an $s-$zero is simply called a zero.

For $j = 1, 2, \cdots, s$, put

$$f_j := \sum_{k=0}^{\mu_j} c_{jk} x_j^k \tag{4}$$

where

$$c_{jk} = c_{jk}(u_1, \cdots, u_d, x_1, \cdots, x_{j-1})$$

are polynomials in $u_1, \cdots, u_d, x_1, \cdots, x_{j-1}$. A zero $(\bar{u}_1, \cdots, \bar{u}_d, \bar{x}_1, \cdots, \bar{x}_s)$ is said to be a *quasi-normal zero* if for every f_j, among the values

$$c_{jk}(\bar{u}_1, \cdots, \bar{u}_d, \bar{x}_1, \cdots, \bar{x}_{j-1}),$$

$$k = 1, 2, \cdots, \mu_j,$$

at least one of them is non-vanishing. In this case we also say that the *weakly non-degenerate condition* holds. A $j-$zero $(\bar{u}_1, \cdots, \bar{u}_d, \bar{x}_1, \cdots, \bar{x}_j)$ is said to be a *quasi-normal $j-$zero* if for $i \leq j$, among the values

$$c_{ik}(\bar{u}_1, \cdots, \bar{u}_d, \bar{x}_1, \cdots, \bar{x}_{i-1}),$$

$$k = 1, 2, \cdots, \mu_i,$$

at least one of them is non-vanishing. A $j-$zero is called a *normal $j-$zero* if it keeps Wu's non-degenerate conition

$$l_1 l_2 \cdots l_j \neq 0. \tag{5}$$

A parameter point $(\bar{u}_1, \cdots, \bar{u}_d)$ is called a *normal $j-$parameter-point* if

$$(\bar{u}_1, \cdots, \bar{u}_d, x_1, \cdots, x_j)$$

is a normal $j-$zero whenever it is a $j-$zero. A normal $s-$parameter-point is simply called a *normal parameter point*.

Without loss of generality, we can assume that every ascending chain depends upon at least one parameter u_1; otherwise, multiply every polynomial of the ascending chain by u_1.

An ascending chain is called a ***proper ascending chain*** if the normal parameter points are dense in the parameter space $\mathbf{C}^d$.

2 Successive Resultant Computation

Some concepts and results on resultant computation will be used later. Given two polynomials, $p(x)$ and $q(x)$,

$$p(x) := a_n x^n + a_{n-1}x^{n-1} + \cdots + a_1 x + a_0,$$

$$q(x) := b_k x^k + b_{k-1}x^{k-1} + \cdots + b_1 x + b_0,$$

by resultant(p, q, x) denote the resultant of p and q with respect to indeterminate x, that is an $(n+k)\times(n+k)$ determinant in terms of the coefficients of $p(x)$ and $q(x)$ as follows:

$$\begin{vmatrix}
a_n & a_{n-1} & \cdots & \cdots & a_0 & & & & \\
 & a_n & a_{n-1} & \cdots & \cdots & a_0 & & & \\
 & & & & \cdots & \cdots & & & \\
 & & & & a_n & a_{n-1} & \cdots & \cdots & a_0 \\
b_k & b_{k-1} & \cdots & \cdots & b_0 & & & & \\
 & b_k & b_{k-1} & \cdots & \cdots & b_0 & & & \\
 & & & & \cdots & \cdots & & & \\
 & & & & b_k & b_{k-1} & \cdots & \cdots & b_0
\end{vmatrix}$$

Given a polynomial $G(u_1, \cdots, u_d, x_1, \cdots, x_s)$ and an ascending chain $\{f_1, \cdots, f_s\}$, putting

$$r_{s-1} := \text{resultant}(G, f_s, x_s),$$

$$r_{s-2} := \text{resultant}(r_{s-1}, f_{s-1}, x_{s-1}),$$

$$r_{s-3} := \text{resultant}(r_{s-2}, f_{s-2}, x_{s-2}),$$

$$\cdots\cdots$$

$$r_1 := \text{resultant}(r_2, f_2, x_2),$$

$$r_0 := \text{resultant}(r_1, f_1, x_1),$$

we call r_0 *the resultant of ascending chain* $\{f_1, \cdots, f_s\}$ *with respect to polynomial* G and denote it simply by

$$\mathrm{res}(f_1, \cdots, f_s, G).$$

Lemma 1. Let $f(x)$ and $g(x)$ be polynomials with degrees μ and ν, roots $x^{(1)}, x^{(2)}, \cdots, x^{(\mu)}$ and $\underline{x}^{(1)}, \underline{x}^{(2)}, \cdots, \underline{x}^{(\nu)}$, and leading coefficients l and $\underline{l}$, respectively. We have

$$\mathrm{resultant}(f, g, x) = l^{\nu} \underline{l}^{\mu} \prod_{i=1}^{\mu} \prod_{j=1}^{\nu} (x^{(i)} - \underline{x}^{(j)}). \tag{6}$$

This well-known fact can be found in algebra textbooks.

Lemma 2. With notations as above,

$$\mathrm{resultant}(f, g, x) = l^{\nu} \prod_{i=1}^{\mu} g(x^{(i)}). \tag{7}$$

This lemma is easy to prove by using Lemma 1.

Now we need more notations. Put $\mu_j := \mathrm{degree}(f_j, x_j)$ for $j = 1, 2, \cdots, s$. Since the system of equations (2) is an ascending chain, theoretically speaking, all indeterminates $x_1, x_2, \cdots, x_s$ can be solved one by one, with several branches each. Thus, we can denote the full solution to (2) by

$$\begin{cases} x_1 = x_1^{(i_1)}(u_1, \cdots, u_d) \\ x_2 = x_2^{(i_1 i_2)}(u_1, \cdots, u_d) \\ x_3 = x_3^{(i_1 i_2 i_3)}(u_1, \cdots, u_d) \\ \cdots\cdots \\ x_s = x_s^{(i_1 i_2 \cdots i_s)}(u_1, \cdots, u_d) \end{cases} \tag{8}$$

where i_j ranges from 1 to μ_j, for $j = 1, 2, \cdots, s$, if the ascending chain is proper.

For simplicity, we put

$$\vec{u} := (u_1, \cdots, u_d),$$

$$\vec{x} := (x_1, \cdots, x_s),$$

and

$$\vec{x}^{(i_1 i_2 \cdots i_s)} := (x_1^{(i_1)}(\vec{u}), x_2^{(i_1 i_2)}(\vec{u}), \cdots, x_s^{(i_1 i_2 \cdots i_s)}(\vec{u})). \tag{9}$$

Lemma 3. Given notations as above, put $\nu_s := \mathrm{degree}(G, x_s)$,

$$\nu_k := \mathrm{degree}(\mathrm{res}(f_{k+1}, f_{k+2}, \cdots, f_s, G), x_k) \tag{10}$$

for $k = 1, 2, \cdots, s-1$, and

$$L(\vec{u}) := l_1(\vec{u})^{\nu_1} \prod_{k=2}^{s} \prod_{i_1=1}^{\mu_1} \prod_{i_2=1}^{\mu_2} \cdots \prod_{i_{k-1}=1}^{\mu_{k-1}} (l_k(\vec{u}, \vec{x}^{(i_1 i_2 \cdots i_s)})^{\nu_k}. \tag{11}$$

where every $l_k(\vec{u}, \vec{x}^{(i_1 \cdots i_s)})$ is an algebraic function of $\vec{u}$ obtained by substituting a branch of the solution, $\vec{x}^{(i_1 \cdots i_s)}$, for the $\vec{x}$ in $l_k(\vec{u}, \vec{x})$, the leading coefficient of f_k with respect to x_k. If $\vec{u}$ is a normal parameter point, then we have

$$\mathrm{res}(f_1, \cdots, f_s, G) = L(\vec{u}) \prod_{i_1=1}^{\mu_1} \prod_{i_2=1}^{\mu_2} \cdots \prod_{i_s=1}^{\mu_s} G(\vec{u}, \vec{x}^{(i_1 i_2 \cdots i_s)}) \tag{12}$$

and $L(\vec{u}) \neq 0$.

It is not difficulty to prove Lamma 3 by using Lemma 2 successively.

Successive resultant computation, which is supported by current softwares of computer algebra such as MAPLE, MATHEMATICA, REDUCE and MACSYMA, has been used in an efficient algorithm introduced by authors[7][8] for searching for the dependency between algebraic equations and applied to automated theorem proving.

3 Main Theorems

In this section, Theorem 1 gives a sufficient and necessary condition for an ascending chain to be a proper ascending chain and Theorem 2 replaces Wu's non-degenerate condition with a much weaker one.

Theorem 1. An ascending chain AS is a proper ascending chain if it has a normal parameter point.

Proof: Do induction on s, the number of polynomials of AS. The conclusion obviously holds for $s = 1$. Assume the conclusion holds for $s = j - 1$. When $s = j$, if the normal parameter points set of AS is not dense, then there is a region **B** in $\mathbf{C}^d$ that doesn't contain any normal parameter point. By AS$'$ denote the ascending Chain which consists of the first $j - 1$ polynomials of AS. Clearly AS$'$ is also a proper ascending chain, so the normal $(j-1)-$parameter points are dense in **B**. Let $R(\vec{u})$ be the resultant of ascending chain $\{f_1, \cdots, f_{j-1}\}$ with respect to l_j. By Lemma 3, for a normal $(j-1)-$parameter-point $\vec{u_0}$, we have

$$R(\vec{u_0}) = L(\vec{u_0}) \prod_{i_1=1}^{\mu_1} \prod_{i_2=1}^{\mu_2} \cdots \prod_{i_{j-1}=1}^{\mu_{j-1}} l_j(\vec{u_0}, \vec{x}^{(i_1 i_2 \cdots i_{j-1})}) \tag{13}$$

where

$$L(\vec{u_0}) := l_1(\vec{u_0})^{\nu_1} \prod_{k=2}^{j-1} \prod_{i_1=1}^{\mu_1} \prod_{i_2=1}^{\mu_2} \cdots \prod_{i_{k-1}=1}^{\mu_{k-1}} (l_k(\vec{u_0}, \vec{x}^{(i_1 i_2 \cdots i_{j-1})})^{\nu_k} \neq 0$$

and every $(\vec{u_0}, \vec{x}^{(i_1 i_2 \cdots i_{j-1})})$ is a normal $(j-1)-$zero.

By the hypothesis, there is a normal parameter point $\vec{u_0}$, for which $R(\vec{u_0}) \neq 0$ because

$$\prod_{i_1=1}^{\mu_1} \prod_{i_2=1}^{\mu_2} \cdots \prod_{i_{j-1}=1}^{\mu_{j-1}} l_j(\vec{u_0}, \vec{x}^{(i_1 i_2 \cdots i_{j-1})}) \neq 0.$$

On the other hand, since **B** contains no normal parameter point, $R(\vec{u}) = 0$ over **B** and over $\mathbf{C}^d$. Contradiction! This completes the proof.

Lemma 4. Given a polynomial over **C**,

$$f(z) := a_n z^n + \cdots + a_k z^k + \cdots + a_1 z + a_0,$$

if $a_n \neq 0, a_k \neq 0, n \geq k \geq 1$ then there is a root z_1 such that

$$|z_1| \leq \sqrt[k]{C_n^k \left|\frac{a_0}{a_k}\right|}. \tag{14}$$

Proof: Let $z_1, z_2, \cdots, z_n$ be the roots of $f(z)$ and

$$|z_1| \leq |z_2| \leq \cdots \leq |z_n|.$$

Then we have

$$C_n^k |z_{k+1} z_{k+2} \cdots z_n| \geq \left|\frac{a_k}{a_n}\right|$$

and

$$|z_1 z_2 \cdots z_n| = \left|\frac{a_0}{a_n}\right|,$$

hence

$$|z_1 z_2 \cdots z_k| \leq C_n^k \left|\frac{a_0}{a_k}\right|.$$

Thus

$$|z_1| \leq \sqrt[k]{C_n^k \left|\frac{a_0}{a_k}\right|}.$$

Lemma 5. Given a polynomial over **C**,

$$g(z) := b_k z^k + \cdots + b_1 z + b_0$$

where $k \geq 1$ and $b_k \neq 0$, an integer $n \geq k$ and a positive number ε, there is a positive number δ such that the ε-neighborhood of every root of $g(z)$ must contain a root of polynomial

$$f(z) := a_n z^n + \cdots + a_1 z + a_0$$

whenever $|a_i| \leq \delta$ for $i = k+1, \cdots, n$ and $|a_j - b_j| \leq \delta$ for $j = 0, 1, \cdots, k$.

Proof: Let z_0 be any root of $g(z)$. By a translation

$$z \mapsto z + z_0,$$

$g(z)$ and $f(z)$ are transformed into

$$\bar{g}(z) = b'_k z^k + b'_{k-1} z^{k-1} + \cdots + b'_1 z \tag{15}$$

and

$$\bar{f}(z) = a'_n z^n + a'_{n-1} z^{n-1} + \cdots + a'_1 z + f(z_0), \tag{16}$$

where

$$b'_j = \sum_{i=j}^{k} b_i C_i^j z_0^{i-j}, \qquad a'_j = \sum_{i=j}^{n} a_i C_i^j z_0^{i-j}.$$

Putting $M := |z_0|^n + |z_0|^{n-1} + \cdots + |z_0| + 1$, we have

$$|f(z_0)| \leq (|z_0|^n + |z_0|^{n-1} + \cdots + |z_0| + 1)\delta = M\delta. \tag{17}$$

Since

$$|a'_k| = |\sum_{i=k}^{n} a_i C_i^k z_0^{i-k}| \geq |a_k| - |\sum_{i=k+1}^{n} a_i C_i^k z_0^{i-k}| \geq |b_k| - \delta - \delta \sum_{i=k+1}^{n} C_i^k |z_0|^{i-k},$$

letting

$$\delta \leq |b_k| / (2(1 + \sum_{i=k+1}^{n} C_i^k |z_0|^{i-k})),$$

we have

$$|a'_k| \geq \frac{1}{2}|b_k|. \tag{18}$$

Combining Lemma 4 with (17) and (18), we see there is at least one root of $\bar{f}(z)$, namely, z_1, such that

$$|z_1| \leq \sqrt[k]{C_n^k \frac{|f(z_0)|}{|a'_k|}} \leq \sqrt[k]{C_n^k \frac{2M\delta}{|b_k|}}.$$

Thus, given $\varepsilon > 0$ and letting

$$\delta \leq \min\{\frac{|b_k|}{2(1 + \sum_{i=k+1}^{n} C_i^k |z_0|^{i-k})}, \frac{\varepsilon^k |b_k|}{2MC_n^k}\},$$

the ε-neighborhood of the zero-root of $\bar{g}(z)$ contains at least one root of $\bar{f}(z)$, that is, the ε-neighborhood of z_0 must contain a root of $f(z)$. This completes the proof.

Theorem 2. Given a proper ascending chain AS $:= \{f_1, \cdots, f_s\}$ and a polynomial g, if every normal zero of AS is a zero of g, then every quasi-normal zero of AS is also a zero of g.

Proof: Otherwise, there is a quasi-normal zero of AS, $(\vec{u}^*, x_1^*, \cdots, x_s^*)$, such that

$$g(\vec{u}^*, x_1^*, \cdots, x_s^*) \neq 0.$$

Then, there is an $\varepsilon > 0$ such that

$$g(\vec{u}, x_1, \cdots, x_s) \neq 0$$

if

$$|\vec{u} - \vec{u}^*| < \varepsilon, |x_1 - x_1^*| < \varepsilon, \cdots, |x_s - x_s^*| < \varepsilon.$$

By Lemma 5, we can find a number δ_s such that, for any $(\vec{u}, x_1, \cdots, x_{s-1})$ satisfying

$$|\vec{u} - \vec{u}^*| < \delta_s, |x_1 - x_1^*| < \delta_s, \cdots, |x_{s-1} - x_{s-1}^*| < \delta_s$$

and

$$l_s(\vec{u}, x_1, \cdots, x_{s-1}) \neq 0,$$

there exists a number x_s, $|x_s - x_s^*| < \varepsilon$ and

$$f_s(\vec{u}, x_1, \cdots, x_s) = 0.$$

Analogously, we can find a number δ_{s-1} such that, for any $(\vec{u}, x_1, \cdots, x_{s-2})$ satisfying

$$|\vec{u} - \vec{u}^*| < \delta_{s-1}, |x_1 - x_1^*| < \delta_{s-1}, \cdots, |x_{s-2} - x_{s-2}^*| < \delta_{s-1}$$

and

$$l_{s-1}(\vec{u}, x_1, \cdots, x_{s-2}) \neq 0,$$

there exists a number x_{s-1}, $|x_{s-1} - x_{s-1}^*| < \min\{\delta_s, \varepsilon\}$ and

$$f_s(\vec{u}, x_1, \cdots, x_{s-1}) = 0.$$

Doing in this way successively, the numbers $\delta_s, \delta_{s-1}, \cdots, \delta_3, \delta_2$ are defined one by one. At last, we can find a number δ_1 such that, whenever $\vec{u}$ satisfies $|\vec{u} - \vec{u}^*| < \delta_1$ and $l_1(\vec{u}) \neq 0$, there exists a number x_1,

$$|x_1 - x_1^*| < \min\{\delta_2, \delta_3, \cdots, \delta_s, \varepsilon\}$$

and $f_1(\vec{u}, x_1) = 0$.
Now we put

$$\delta := \min\{\delta_1, \delta_2, \cdots, \delta_s, \varepsilon\}.$$

The normal parameter points are dense in $\mathbf{C}^d$ because AS is a proper ascending chain, so there is a normal parameter point $\vec{u_0}$ satisfying $|\vec{u_0} - \vec{u}^*| < \delta$. Starting from $\vec{u_0}$, by the above argument, we can find successively the numbers $x_1^0, x_2^0, \cdots, x_s^0$ such that

$$(\vec{u_0}, x_1^0, \cdots, x_s^0)$$

is a zero of AS and $|x_j^0 - x_j^*| < \varepsilon$ for $j = 1, 2, \cdots, s$.
Clearly, this is a normal zero of AS since $\vec{u_0}$ is chosen to be a normal parameter point, so it is also a zero of g. However, as we have pointed out at the beginning of the proof, there is no zero of g in such a neighborhood. It is a contradiction. This completes the proof.

Denote by ZERO(.) the set of zeros of a polynomial or a system of polynomials. The following corollaries are obvious:

Corollary 1. Let $\{f_1, \cdots, f_s\}$ be a proper ascending chain and c_{jk} the coefficients of f_j in x_j, for $j = 1, 2, \cdots, s$ and $k = 0, 1, \cdots, \mu_j$, as defined in (4). We have

$$\mathrm{ZERO(AS)} - \bigcup_{j=1}^{s} \left(\bigcap_{k=1}^{\mu_j} \mathrm{ZERO}(c_{jk})\right) \subset \mathrm{ZERO(PS)}.$$

Corollary 2. On proving an equality-type theorem, if the final pseudo remainder of the conclusion polynomial with respect to the proper ascending chain which is obtained by well-ordering from the hypothesis is identical to zero, then the conclusion holds under the weakly non-degenerate condition.

Acknowledgments

One of the authors (Y.L.) would like to thank Professor Abdus Salam, the International Atomic Energy and UNESCO for hospitality at the International Centre for Theoretical Physics, Trieste.

References

[1] Chou, S. C., *Mechanical Geometry Theorem Proving,* D. Reidel Publishing Company, (Amsterdam) 1988.

[2] Chou, S. C. and Gao X. S., *A class of geometry statements of constructive type and geometry theorem proving,* Technical Report: TR-89-37, Dept. of Computer Sciences, Univ. of Texas at Austin, 1989.

[3] Kutzler, B., 'Careful algebraic translations of geometry theorems', in *Proceedings of the 1989 Symposium on Symbolic and Algebraic Computation,* 254-263.

[4] Ritt, J. F., *Differential Equations from the Algebraic Standpoint,* Amer. Math. Soc., 1932.

[5] Ritt, J. F., *Differential Algebra,* Amer. Math. Soc., 1950.

[6] Wu Wen-tsün, 'On the decision problem and the mechanization of theorem proving in elementary geometry', *Scientia Sinica* **21**, 157-179, (1978).

[7] Yang Lu and Zhang Jingzhong, 'Searching dependency between algebraic equations: An algorithm applied to automated reasoning', in *Artificial Intelligence in Mathematics,* IMA Conference Proceedings, Oxford University Press, to appear.

[8] Zhang Jingzhong, Yang Lu & Hou Xiaorong, 'A criterion for dependency of algebraic equations and its application to automated theorem proving', *Proc. of 1992 IWMM,* Beijing, International Academic Publishing Company.

Computer Mathematics
Proc. of the Special Program at
Nankai Institute of Mathematics
January 1991 – June 1991

Mechanical Theorem Proving in Riemann Geometry Using Wu's Method*

Shang-Ching Chou
Department of Computer Science
The Wichita State University
Wichita, KS 67260, USA

Xiao-Shan Gao
Institute of Systems Science
Academia Sinica
Beijing 100080, P.R. China

Abstract. We report some results about the mechanical theorem proving in Riemann geometry. We prove a theorem which reduces a geometry statement in Riemann geometry to several substatements which are much easier to prove. For a class of constructive geometry statements, we present a method to generate sufficient non-degenerate conditions in geometric form as well as a method to prove whether the statements are true under the mechanically generated conditions. We also prove that an irreducible constructive statement is generally true if and only if it is true under the non-degenerate conditions generated by our method.

1. Introduction

In [WU1], Wu Wen-tsün introduced an algebraic method which can be used to prove quite non-trivial theorems of equation type. Hundreds of theorems from Euclidean geometry have been proved with Wu's method [WU4, CH2, WG1]. Quite a few theorems from various non-Euclidean geometries have also been proved by the method [WU4 , CK1, GA1]. Inspired by Wu's work, people also presented methods based on the Gröbner basis method to prove the same class of geometry theorems which Wu's

* The work reported here was supported in part by the NFS of China and by the NSF (USA) Grant CCR-917870.

method addresses [CH2, KA1, KU1].

From the theoretic point of view, Wu's method can also be used to prove theorems in other elementary geometries including Riemann geometry, as showed by Wu in [WU2]. But to develop an efficient and comparatively complete prover for a special geometry, a lot of detailed work still needs to be done. In this paper, we prove a theorem which can be used to reduce the proof of a geometry statement in Riemann geometry to some special cases which are much easier to prove. We also present a method to generate sufficient non-degenerate conditions in geometric form for a class of constructive geometry statements. These results make the mechanical theorem proving in Riemann geometry much clearer and easier. By a transformation theorem proved in [GA1], the results obtained in this paper for Riemann geometry are also true in Bolyai-Lobachevsky geometry, co-Bolyai-Lobachevsky geometry, and doubly hyperbolic geometry.

We adopt a model for Riemann geometry. In this model the algebraic translation of a geometry statement involving segment congruence is always reducible and the proof of the statement can be reduced to many subcases. We prove that it suffices to check some of the subcases. In most of the examples encountered, we only need to check one of the subcases.

In the usual description of a geometry statement, necessary non-degenerate conditions for the statement to be true are usually not given explicitly and some of them are not easy to find. There are two approaches to dealing with these implicit non-degenerate conditions in mechanical theorem proving using algebraic methods. The first approach is to prove a statement to be generally true at the same time give certain non-degenerate conditions in algebraic form under which the statement is true. The second approach is to prove a statement to be true under certain non-degenerate conditions given explicitly as a part of the geometry statement. A detailed discussion of the formulation problem can be found in [CY1]. As mentioned above, the first approach can generate non-degenerate conditions automatically, but there is no general method to transform these conditions to geometric form. The second approach actually needs people to find the non-degenerate conditions. But generally, it is not easy to find the sufficient non-degenerate conditions for some statements. So it is important that we can generate non-degenerate conditions in geometric form automatically, such that the statement is true in the usual sense iff it is true under these non-degenerate conditions. If certain non-degenerate conditions satisfy this condition, we say they are sufficient (for the precise meaning, see Section 4). In his thesis, the first author has done this for a class of constructive geometry statements [CH1]. In [CG2], we prove that for a small class of geometry statements, the non-degenerate conditions generated by our method are also sufficient for a statement to be true in Euclidean geometry.

In this paper, we extend the results in [CH1,CG2] to Riemann geometry: we describe a class of geometry statements of constructive type, and for statements in this class we present a method of generating sufficient non-degenerate conditions in geometric form for the statement as well as a method to decide whether the statement is true under the automatically generated conditions. For a class of irreducible statements, we prove a theorem to connect the two approaches to dealing with non-degenerate conditions, i.e., an irreducible statement is generally true iff it is true under the geometric non-degenerate conditions generated by our method.

Section 2 discusses the general case of mechanical theorem proving in Riemann geometry. Section 3 presents a class of constructive statements. Section 4 proves the completeness of the non-degenerate conditions. Section 5 provides two examples to illustrate our method.

2. Mechanical Theorem Proving in Riemann Geometry

2.1. A Model for Riemann Geometry

Let $\langle x, y\rangle$ and $x \times y$ be the inner product and vector product in $\mathbf{R}^3$ for x, y in $\mathbf{R}^3$. We denote $\|x\| = \langle x, x\rangle$ and $|x| = \sqrt{\langle x, x\rangle}$ for $x \in \mathbf{R}^3$. Let **S2** be the unit sphere in $\mathbf{R}^3$. Regarding a pair of antipodal points of **S2** as the same point, we get the real projective plane: $\mathbf{P2} = \{\{X, -X\}/X \in \mathbf{S2}\}$ which is a model for Riemann geometry.

Let $\pi : \mathbf{S2} \longrightarrow \mathbf{P2}$ be the mapping that sends each $X \in \mathbf{S2}$ to $\{X, -X\} \in \mathbf{P2}$. If $A = \pi(x, y, z)$, we say (x, y, z) is a *coordinate* for A. A point A on **P2** has two coordinates (x, y, z) and $(-x, -y, -z)$. One of them is called the *antipodal point* of the other. Two points $A = \pi(x_1, y_1, z_1)$ and $B = \pi(x_2, y_2, z_2)$ on **P2** are equal iff $(x_1, y_1, z_1) = (x_2, y_2, z_2)$ or $(x_1, y_1, z_1) = -(x_2, y_2, z_2)$, or equivalently iff $(x_1, y_1, z_1) \times (x_2, y_2, z_2) = 0$.

A line in **P2** is a set of the form $\pi(l)$, where l is the intersection of **S2** and a plane passing the origin. Thus, the equation of a line in **P2** can be expressed as

$$u_1x_1 + u_2x_2 + u_3x_3 = 0$$

where $U = (u_1, u_2, u_3) \neq 0$ is the normal vector of the line and $X = \pi(x_1, x_2, x_3)$ is an arbitrary point on the line. Two lines are equal if their normal vectors are parallel. Two lines are perpendicular if the inner product of their normals is zero. Note that a point P in **P2** determines a unique line l in **P2** whose normal vector is a coordinate of P. P is called the *pole* of l and l is called the *polar* of P.

We define the length of segment XY as

$$D(X, Y) = \arccos(|\langle x, y\rangle|)$$

where $X = \pi(x)$ and $Y = \pi(y)$. We assume that all distances are $\leq \pi/2$. The definition is well-defined for the choices of coordinates of X and Y.

Let $A = \pi(a), B = \pi(b), C = \pi(c), X = \pi(x)$ be points in **P2**. We consider four kinds of straight lines in **P2**.

(1) $l = P(A)$ is the polar line of point A on **P2**. The equation of line $P(A)$ is

$$\langle a, x\rangle = 0$$

(2) $l = L(AB)$ is the line joins points A and B in **P2**. The equation of line $L(AB)$ is

$$\langle a \times b, x\rangle = 0$$

(3) $l = T(C, AB)$ is the line passing through C and is perpendicular to line $L(AB)$ for points A, B, and C in **P2**. The equation of line $T(C, AB)$ is

$$\langle c \times (a \times b), X\rangle = 0$$

(4) $l = B(AB)$ is the perpendicular-bisector of segment AB for points A and B in **P2**. The equation of line $B(AB)$ is

$$\langle a - b, x\rangle\langle a + b, x\rangle = 0$$

Note that $B(AB)$ consists of two lines. It is not difficult to verify that the lines are well defined.

A circle h in **P2** is a pair of a point O and a segment AB: $h = (O, AB)$ which represents the set of points the distances between which and O equal the length of AB, i.e., h is the circle with O as the center and segment AB as the radius. Two circles are equal if their centers are equal and their radii have the same length.

Let Π be a set of points on **P2**. Then a line or a circle is said in Π if the points occurring in the definition of the line or the circle are in Π.

We define the measure of the angle formed by line $L(AB)$ and line $L(CD)$ as

$$\angle(AB, CD) = \arccos(|\langle \frac{a \times b}{|a \times b|}, \frac{d \times c}{|d \times c|}\rangle|)$$

where $\pi(a) = A, \pi(b) = B$, $\pi(c) = C$, $\pi(d) = D$. We assume that all angles are $\leq \pi/2$.

For more details of this model, see [RY1].

2.2. Basic Predicates and Mechanical Theorem Proving

Let $A = \pi(a)$, $B = \pi(b)$, $C = \pi(c)$, $D = \pi(d)$, $X = \pi(x)$, $Y = \pi(y)$, and $Z = \pi(z)$ be points in **P2**. We define the following predicates*.

(1) Polar(A, B) means A is on the polar line of B. Its algebraic equation is

---

* Strictly speaking, they are geometric relations as they can be deduced from the definition of the model.

$$\langle A, B\rangle = 0.$$

(2) Coll(A, B, C) means A, B, and C are on the same line. Its algebraic equation is

$$\langle a, b \times c\rangle = 0.$$

(3) Perp(A, B, C, D) means that $A = B$, or $C = D$, or line $L(AB)$ is perpendicular to line $L(CD)$. Its algebraic equation is

$$\langle a \times b, c \times d\rangle = 0.$$

(4) Cong(A, B, C, D) means the length of AB equals to the length of CD. Its algebraic equation is

$$\langle a, b\rangle^2 = \langle c, d\rangle^2.$$

(5) Acong$(A, B, C; X, Y, Z)$ means that angle XYZ is congruent to angle ABC. Its algebraic equation is

$$\frac{\langle z \times y, z \times y\rangle^2}{\|z \times y\| \cdot \|z \times y\|} = \frac{\langle a \times b, c \times b\rangle^2}{\|a \times b\| \cdot \|c \times b\|}.$$

(6) Pole(A, B, C) means $B = C$ or A is the pole of $L(BC)$. Its algebraic equation is

$$a \times (b \times c) = 0.$$

(7) Para(A, B, C, D) means $A = B$ or $C = D$ or A, B, C, and D are on the same line. Its algebraic equation is

$$(a \times b) \times (c \times d) = 0.$$

(8) Cperp(A, B, C, X, Y) means that para(B, C, X, Y) or A is on the co-perpendicular line of $L(BC)$ and $L(XY)$. Its algebraic equation is

$$(a \times (b \times c)) \times (a \times (x \times y)) = 0.$$

For a predicate P, let $E(P)$ be the set of polynomials representing P.

Definition 2.1. A geometry statement of equation type (or simply a geometry statement) in Riemann geometry is a triple (HS, DS, G) where $HS = \{P_1, ..., P_k\}$ and $DS = \{Q_1, ..., Q_m\}$ for predicates P_i and Q_i, and G is a predicate. HS is called the *equation part* of the hypothesis. DS is called the *inequation part* of the hypothesis.

Definition 2.2. (a) An algebraic statement of equation type (or simply an algebraic statement) is a triple (ES, IS, C) where $ES = \{H_1, ..., H_s\}$, $IS = \{D_1, ..., D_t\}$ are polynomial sets, and C is a polynomial.

(b) An algebraic statement (ES, IS, C) is true in the real number case if

$$\forall x \in \mathbf{R}((H_1 = 0 \wedge ... \wedge H_s = 0 \wedge D_1 \neq 0 \wedge ... \wedge D_t \neq 0) \Rightarrow C = 0)$$

where the x are the variables occurring in H_i, D_i, C, and $\mathbf{R}$ is the real number field.

(c) An algebraic statement (ES, IS, C) is universally true if

$$\forall x \in \mathbf{C}((H_1 = 0 \wedge ... \wedge H_s = 0 \wedge D_1 \neq 0 \wedge ... \wedge D_t \neq 0) \Rightarrow C = 0)$$

where $\mathbf{C}$ is the complex number field.

If a statement is universally true, then it is also true in the real number case. We have.

Theorem 2.3. We can decide whether an algebraic statement is universally true using Ritt-Wu's decomposition algorithm.

Proof. See for example [CG1]. ∎

Let $S = (HS, DS, G)$ ($HS = \{P_1, ..., P_k\}$ and $DS = \{Q_1, ..., Q_m\}$) be a geometry statement. For points occurring in (HS, DS, G), we assign coordinates such that no coordinates of different points can be the same. Then S can be transformed to an algebraic statement $S' = (HS', DS', E(G))$ where $HS' = \bigcup_{1\leq i\leq k} E(P_i)$ and $DS' = \{E(Q_1), ..., E(Q_m)\}$*. $(HS', DS', E(G))$ is called the *algebraic version* of (HS, DS, G). We say S is true if S' is true over the field of real numbers. We say S is *universally* true if S' is universally true. By Theorem 2.3, we can decide whether a geometry statement is universally true.

As mentioned in the introduction, there is another approach of mechanical theorem proving. First, we have the following definition.

Definition 2.4. For a polynomial set $ES = \{H_1, ..., H_s\}$ and a polynomial C, we say (ES, C) or

$$\forall x((H_1 = 0 \wedge \cdots \wedge H_s = 0) \Rightarrow C = 0)$$

is generally true wrpt a set of variables $u_1, ..., u_q$, if there is a polynomial D of the u such that $Zero(ES) \subset Zero(CD)$.

A detailed discussion of the concept of generally true can be found in [WU2,CY2]. We have

Theorem 2.5. Given a polynomial set ES, a polynomial C, and a set of variables $u_1, ..., u_p$, we can decide whether (ES, C) is generally true wrpt $u_1, ..., u_p$ using Ritt-Wu's decomposition algorithm.

Proof. See [WU2,CH2]. ∎

For a geometry statement $S = (HS, DS, G)$, let $S' = (HS', DS', E(G))$ be its algebraic version. We can divide the variables occurring in HS' and $E(G)$ into two

---

* When $E(Q_i) = \{f_1, ..., f_p\}$ contains more than one polynomials, we can also use the following trick to transform $\neg(E(Q_i) = 0)$ to an inequation of a single polynomial: $\exists z(z_1 f_1 + \cdots + z_p f_p \neq 0)$ for new variables z_i.

groups: $u_1,\ldots,u_p$ and $x_1,\ldots,x_q$ such that the variables $u_1,\ldots,u_p$ can take arbitrary value and once their values are fixed the x can be generally determined by the hypothesis HS'. We call the u the parameters of the statement. We define that (HS,DS,G) is generally true wrpt the u if $(HS',E(G))$ is generally true wrpt the u. By Theorem 2.5, we can decide whether a geometry statement with empty inequation part is generally true.

2.3. The Reducibility Problem

Note that the algebraic equation for the predicate cong is always reducible. A statement involving this predicate can be reduced into many subcases.

Example 2.6. In any triangle the three perpendicular-bisectors are concurrent.

Let $A=\pi(0,0,1)$, $B=\pi(0,x_1,x_2)$, $C=\pi(x_3,x_4,x_5)$, and $O=\pi(x_6,x_7,x_8)$.

The hypotheses are

$$h_1 = x_2^2 + x_1^2 - 1 = 0 \qquad B \in \mathbf{P2}.$$
$$h_2 = x_5^2 + x_4^2 + x_3^2 - 1 = 0 \qquad C \in \mathbf{P2}.$$
$$h_3 = x_8^2 + x_7^2 + x_6^2 - 1 = 0 \qquad O \in \mathbf{P2}.$$
$$h_4 = x_8^2 - (x_1x_7 + x_2x_8)^2 = 0 \qquad \mathrm{cong}(O,A,O,B).$$
$$h_5 = (x_6x_3 + x_7x_4 + x_8x_5)^2 - x_8^2 = 0 \qquad \mathrm{cong}(O,A,O,C).$$

The conclusion is

$$c = (x_7x_1 + x_8x_2)^2 - (x_6x_3 + x_7x_4 + x_8x_5)^2 = 0 \qquad \mathrm{cong}(O,B,O,C).$$

Note that $h_4 = h_4'h_4''$ and $h_5 = h_5'h_5''$, where $h_4' = x_8 - (x_1x_7 + x_2x_8)$, $h_4'' = x_8 + (x_1x_7 + x_2x_8)$, $h_5' = x_6x_3 + x_7x_4 + x_8x_5 - x_8$, and $h_5'' = x_6x_3 + x_7x_4 + x_8x_5 + x_8$. Then the above geometry statement is true if and only if

$$\forall x((h_1 = 0 \wedge h_2 = 0 \wedge h_3 = 0 \wedge h_4' = 0 \wedge h_5' = 0) \Rightarrow c = 0)$$
$$\forall x((h_1 = 0 \wedge h_2 = 0 \wedge h_3 = 0 \wedge h_4'' = 0 \wedge h_5' = 0) \Rightarrow c = 0)$$
$$\forall x((h_1 = 0 \wedge h_2 = 0 \wedge h_3 = 0 \wedge h_4' = 0 \wedge h_5'' = 0) \Rightarrow c = 0)$$
$$\forall x((h_1 = 0 \wedge h_2 = 0 \wedge h_3 = 0 \wedge h_4'' = 0 \wedge h_5'' = 0) \Rightarrow c = 0)$$

are true. To deal with this problem, for $A=\pi(x_1,y_1,z_1)$, $B=\pi(x_2,y_2,z_2)$, $C=\pi(x_3,y_3,z_3)$, and $D=\pi(x_4,y_4,z_4)$ we introduce two new predicates:

(1) cong1(A,B,C,D) means $x_1x_2 + y_1y_2 + z_1z_2 = x_3x_4 + y_3y_4 + z_3z_4$.
(2) cong2(A,B,C,D) means $x_1x_2 + y_1y_2 + z_1z_2 = -(x_3x_4 + y_3y_4 + z_3z_4)$.

A *substatement* of a geometry statement $S=(HS,DS,G)$ is a statement $S'=(\,HS,$ $DS,\,G\,)$ where HS' is obtained by replacing the predicates cong in HS by cong1 or cong2. If there are m predicates cong occurring in HS, then S can be reduced to 2^m substatements. We actually do not need to check all of these substatements. In the following, we prove a theorem which can be used to reduce the number of substatements needed to check drastically.

If point A occurs one or three times in cong1(A,B,C,D), then cong1(A,B,C,D) changes to cong2(A,B,C,D) when replacing a coordinate of A with its antipodal. If point A occurs two or four times in cong1(A,B,C,D), then cong1(A,B,C,D) changes to itself when replacing a coordinate of A with its antipodal. Thus we can define an equivalent relation among the substatements: two substatements of a geometry statement are *equivalent* if one of them can be changed to the other by replacing the coordinates for some points in the predicates by their antipodals.

Theorem 2.7. Let (HS', DS, G) and (HS'', DS, G) be two equivalent substatements of a statement (HS, DS, G), then (HS', DS, G) is (universally) true if and only if (HS'', DS, G) is (universally) true.

Proof. Let $HS' = \{P_1, ..., P_k\}$, $DS = \{D_1, ..., D_l\}$. Then the substatement (HS', DS, G) is true iff

(2.8) $\forall x \in \mathbf{R}(E(P_1) = 0 \wedge ... \wedge E(P_k) = 0 \wedge E(D_1) \neq 0 \wedge ... \wedge E(D_l) \neq 0 \Rightarrow E(G) = 0)$

Let $HS'' = \{P'_1, ..., P'_k\}$. The substatement (HS'', DS, G) is true iff

(2.9) $\forall x \in \mathbf{R}(E(P'_1) = 0 \wedge ... \wedge E(P'_k) = 0 \wedge E(D_1) \neq 0 \wedge ... \wedge E(D_l) \neq 0 \Rightarrow E(G) = 0)$

As the two substatements are equivalent, $E(P_i) = E(P'_i)$ for $i = 1, ..., k$ when replacing the coordinates for some points by their antipodals in one substatement, i.e., when replacing some variables x_i by $-x_i$, (2.8) becomes (2.9). Thus (2.8) is true if and only if (2.9) is true. By changing the $\mathbf{R}$ in (2.8) and (2.9) to $\mathbf{C}$, we can get the result about the universally true. ∎

Theorem 2.10. In a statement (HS, DS, G), if for each predicate cong(P_1, P_2, P_3, P_4) in HS there is a point P_i which occurs in this predicate one or three times and does not occur in other cong predicates in HS, then the statement (HS, DS, G) is true iff one of its substatements is true.

Proof. Under the condition of this theorem, each predicate cong1 (cong2) in a substatement of (HS, DS, G) can be changed to cong2 (cong1) separately. Thus all the substatements are equivalent. Now the theorem comes from Theorem 2.7. ∎

By Theorem 2.10, for Example 2.6 we only need to check one of its substatements as point B occurs in cong(O, A, O, B) only and point C occurs in cong(O, A, O, C) only.

3. A Class of Geometry Statements of Constructive Type

3.1. The Constructions

Definition 3.1. A construction is one of the six operations.

Construction 1. Take an arbitrary point P in **P2**.
Construction 2. Take an arbitrary point P on a line l in **P2**.
Construction 3. Take an arbitrary point P on a circle h in **P2**.
Construction 4. Take the intersection P of two lines l_1 and l_2 in **P2**.
Construction 5. Take an intersection P of a line l and a circle h in **P2**.
Construction 6. Take an intersection P of two circles h_1 and h_2 in **P2**.

The point P in each of the above constructions is said to *be introduced by the construction*. In the following, we shall give the exact geometric meaning in terms of geometry predicates and the algebraic translation for each of the above constructions. The algebraic translation of a construction consists of two parts: the *equation part* HS and the *inequation part* DS. For any point P in **P2**, we use its lowercase p to represent a point in **S2** such that $\pi(p) = P$ and $\|p\| = 1$.

Construction 1. Take an arbitrary point P.

$$HS = \emptyset, \quad DS = \emptyset.$$

Construction 2. Take an arbitrary point P on a line l. We have four cases as there exist four kinds of lines.

Case 2.1. $l = P(P_1)$.

$$HS = \{\text{polar}(P, P_1)\}, \quad DS = \{\}$$

Case 2.2. $l = L(P_1P_2)$.

$$HS = \{\text{coll}(P, P_1, P_2)\}, \quad DS = \{\|p_1 \times p_2\| \neq 0\}$$

In the real number case DS is equivalent to $P_1 \neq P_2$.*

Case 2.3. $l = T(P_3, P_1P_2)$.

$$HS = \{\text{perp}(P, P_3, P_1, P_2)\}, \quad DS = \{\|p_3 \times (p_2 \times p_1)\| \neq 0\}$$

In the real number case DS is equivalent to $\neg\text{pole}(P_3, P_1, P_2)$.

Case 2.4. $l = B(P_1P_2)$.

$$HS = \{\text{cong}(P, P_1, P, P_2)\}, \quad DS = \{\|p_2 - p_1\|\|p_2 + p_1\| \neq 0\}$$

In the real number case DS is equivalent to $P_1 \neq P_2$.

Construction 3. Take an arbitrary point P on a circle $h = (O, AB)$.

$$HS = \{\text{cong}(O, P, A, B)\}, \quad DS = \emptyset$$

Construction 4. Take the intersection P of two lines l_1 and l_2. Generally speaking the non-degenerate condition is that l_1 is not the same as l_2.

* The proof of this result can be found in Appendix A. The same for the following constructions.

Case 4.1. $l_1 = P(P_1)$ and $l_2 = P(P_2)$.

$HS = \{\mathrm{polar}(P, P_1), \mathrm{polar}(P, P_2)\}, \quad DS = \{\|p_1 \times p_2\| \neq 0\}$

In the real number case DS is equivalent to $P_1 \neq P_2$.

Case 4.2. $l_1 = P(P_1)$ and $l_2 = L(P_2, P_3)$.

$HS = \{\mathrm{polar}(P, P_1), \mathrm{coll}(P, P_2, P_3)\}, \quad DS = \{\|p_1 \times (p_2 \times p_3)\| \neq 0\}$

In the real number case DS is equivalent to $\neg\mathrm{pole}(P_1, P_2, P_3)$.

Case 4.3. $l_1 = P(P_1)$ and $l_2 = T(P_2, P_3, P_4)$.

$HS = \{\mathrm{polar}(P, P_1), \mathrm{perp}(P, P_2, P_3, P_4)\}, \quad DS = \{\|p_1 \times (p_2 \times (p_3 \times p_4))\| \neq 0\}$

In the real number case DS is equivalent to $\neg\mathrm{pole}(P_2, P_3, P_4)$ and $(\neg\mathrm{polar}(P_1, P_2)$ or $\neg\mathrm{coll}(P_1, P_3, P_4))$.

Case 4.4. $l_1 = P(P_1)$ and $l_2 = B(P_2, P_3)$.

$HS = \{\mathrm{polar}(P, P_1), \mathrm{cong}(P, P_2, P, P_3)\}, \quad DS = \{\|p_1 \times (p_2 \pm p_3)\| \neq 0\}$

In the real number case DS is equivalent to
$\neg\mathrm{coll}(P_1, P_2, P_3)$ or $\neg\mathrm{cong}(P_1, P_2, P_1, P_3)$ (P_1 is not the middle point of P_2P_3).

Case 4.5. $l_1 = L(P_1P_2)$ and $l_2 = L(P_3P_4)$.

$HS = \{\mathrm{coll}(P, P_1, P_2), \mathrm{coll}(P, P_3, P_4)\}, \quad DS = \{\|(p_1 \times p_2) \times (p_3 \times p_4)\| \neq 0\}$

In the real number case DS is equivalent to $\neg\mathrm{para}(P_1, P_2, P_3, P_4)$.

Case 4.6. $l_1 = L(P_1P_2)$ and $l_2 = T(P_3, P_4P_5)$.

$HS = \{\mathrm{coll}(P, P_1, P_2), \mathrm{perp}(P, P_3, P_4, P_5)\}, \quad DS = \{\|(p_1 \times p_2) \times (p_3 \times (p_4 \times p_5))\| \neq 0\}$

In the real number case DS is equivalent to $\neg\mathrm{pole}(P_3, P_4, P_5)$ and $(\neg\mathrm{coll}(P_1, P_2, P_3)$ or $\neg\mathrm{perp}(P_1, P_2, P_4, P_5))$.

case 4.6.1. $l_1 = L(P_1P_2)$ and $l_2 = T(P_3, P_1P_2)$. (The foot from P_3 to $L(P_1, P_2)$.)

$HS = \{\mathrm{coll}(P, P_1, P_2), \mathrm{perp}(P, P_3, P_1, P_2)\}, DS = \{\|(p_1 \times p_2) \times (p_3 \times (p_1 \times p_2))\| \neq 0\}$

In the real number case DS is equivalent to $\neg\mathrm{pole}(P_3, P_1, P_2)$.

Case 4.7. $l_1 = L(P_1P_2)$ and $l_2 = B(P_3P_4)$.

$HS = \{\mathrm{coll}(P, P_1, P_2), \mathrm{cong}(P, P_3, P, P_4)\}, \quad DS = \{\|(p_1 \times p_2) \times (p_4 \pm p_3)\| \neq 0\}$

In the real number case DS is equivalent to
$P_1 \neq P_2$ and $(\neg\mathrm{cong}(P_1, P_3, P_1, P_4)$ or $\neg\mathrm{perp}(P_1, P_2, P_3, P_4))$.

case 4.7.1. $l_1 = L(P_1P_2)$ and $l_2 = B(P_1P_2)$. (The middle point of P_1P_2.)

$HS = \{\mathrm{coll}(P, P_1, P_2), \mathrm{cong}(P, P_1, P, P_2)\}, \quad DS = \{\|(p_1 \times p_2) \times (p_1 \pm p_2)\| \neq 0\}$

In the real number case DS is equivalent to $P_1 \neq P_2$.

Case 4.8. $l_1 = T(P_1, P_2P_3)$ and $l_2 = T(P_4, P_5P_6)$.

$HS = \{\text{perp}(P, P_1, P_2, P_3), \text{perp}(P, P_4, P_5, P_6)\}$
$DS = \{\|(p_1 \times (p_2 \times p_3)) \times (p_4 \times (p_5 \times p_6))\| \neq 0\}$

In the real number case DS is equivalent to
$\neg$pole(P_1, P_2, P_3) and $\neg$pole(P_4, P_5, P_6) and ($P_1 \neq P_4$ or $\neg$cperp$(P_1, P_2, P_3, P_5, P_6)$) and ($\neg$perp$(P_1, P_4, P_2, P_3)$ or $\neg$perp(P_1, P_4, P_5, P_6)).

Case 4.9. $l_1 = T(P_1, P_2P_3)$ and $l_2 = B(P_4P_5)$.

$HS = \{\text{perp}(P, P_1, P_2, P_3), \text{cong}(P, P_4, P, P_5)\}$, $DS = \{\|(p_1 \times (p_2 \times p_3)) \times (p_5 \pm p_4)\| \neq 0\}$

In the real number case DS is equivalent to
$\neg$pole(P_1, P_2, P_3) and ($\neg$cong(P_1, P_4, P_1, P_5) or $\neg$cperp$(P_1, P_2, P_3, P_4, P_5)$).

Case 4.10. $l_1 = B(P_1P_2)$ and $l_2 = B(P_3P_4)$.

$HS = \{\text{cong}(P, P_1, P, P_2), \text{cong}(P, P_3, P, P_4)\}$, $DS = \{\|(p_2 \pm p_1) \times (p_4 \pm p_3)\| \neq 0\}$

In the real number case DS is equivalent to $P_1 \neq P_2$ and $P_3 \neq P_4$ and ($\neg$cong(P_1, P_3, P_2, P_4) or $\neg$cong(P_1, P_4, P_2, P_3)).

Construction 5. Take an intersection P of a line l and a circle h.

Case 5.1. $l = P(P_1)$ and $h = (O, Q_1Q_2)$.

$HS = \{\text{polar}(P, P_1), \text{cong}(P, O, Q_1, Q_2)\}$, $DS = \{\|p_1 \times o\| \neq 0\}$

In the real number case DS is equivalent to $P_1 \neq O$.

Case 5.2. $l = L(P_1P_2)$ and $h = (O, Q_1Q_2)$.

$HS = \{\text{coll}(P, P_1, P_2), \text{cong}(P, O, Q_1, Q_2)\}$, $DS = \{\|o \times (p_1 \times p_2)\| \neq 0\}$

In the real number case DS is equivalent to $\neg$pole(O, P_1, P_2).

Case 5.3. $l = T(P_1, P_2P_3)$ and $h = (O, Q_1Q_2)$.

$HS = \{\text{perp}(P, P_1, P_2, P_3), \text{cong}(P, O, Q_1, Q_2)\}$, $DS = \{\|(p_1 \times (p_2 \times p_3)) \times o\| \neq 0\}$

In the real number case DS is equivalent to $\neg$pole(P_1, P_2, P_3) and ($\neg$polar(P_1, O) or $\neg$coll(O, P_2, P_3)).

Case 5.4. $l = B(P_1P_2)$ and $h = (O, Q_1Q_2)$.

$HS = \{\text{cong}(P, O, Q_1, Q_2), \text{cong}(P, P_1, P, P_2)\}$, $DS = \{\|(p_2 \pm p_1) \times o\| \neq 0\}$

In the real number case DS is equivalent to
$\neg$coll(O, P_1P_2) or $\neg$cong(O, P_1, O, P_2) (O is not the middle point of P_1P_2).

Construction 6. Take an intersection P of two circles $h_1 = (O_1, P_1P_2)$ and $h_2 =$

(O_2, Q_1Q_2).

$HS = \{\text{cong}(P, O_2, Q_1, Q_2), \text{cong}(P, O_2, P_1, P_2)\}$, $DS = \{\|o_1 \times o_2\| \neq 0\}$

In the real number case DS is equivalent to $O_1 \neq O_2$.

The equation part HS of a construction is in geometric form, but its inequation part DS is in algebraic form and can be transformed into geometry predicates in the real number case.

3.2. Geometry Statements of Constructive Type

Definition 3.2. A construction sequence is a sequence of constructions such that for each construction in the sequence the point introduced by it must be different from the points introduced by the previous constructions and the lines and circles occurring in this construction must be in the set of points introduced by the previous constructions.

Definition 3.3. A geometry statement of constructive type is a pair (CS, G) where CS is a construction sequence and G is a predicate.

For a statement of constructive type (CS, G), we give different and non-zero coordinates for different points. Then (CS, G) can be transformed to an algebraic statement $(ES, IS, E(G))$ where $ES = \{H_1, ..., H_s\}$ is the set of polynomials in the equation part of CS; $IS = \{D_1, ..., D_t\}$ is the set of polynomials in the inequation part of CS. We call $(ES, IS, E(G))$ the *algebraic version* of (CS, G).

Definition 3.3. Let $(ES, IS, E(G))$ be the algebraic version of a constructive statement (CS, G). We say (CS, G) is true in Riemann geometry if $(ES, IS, E(G))$ is true in the real number case and (CS, G) is universally true if $(ES, IS, E(G))$ is universally true.

By Theorems 2.3 and 2.5, we can decide whether a constructive statement is universally true or generally true.

In the real number case, we can represent a statement of constructive type by geometry predicates. Let $HS = \{P_1, ..., P_s\}$ be the predicates in the equation part of the statement, and $DS = \{Q_1, ..., Q_t\}$ be the predicate formulas which are equivalent to the inequation part of the statement in the real number case. Then the statement (CS, G) is equivalent to the geometry statement (HS, DS, G) in the real number case.

4. The Completeness of the Non-degenerate Conditions

In this section, we shall prove that the non-degenerate conditions generated in Section 3.1 for a statement of constructive type (CS, G) are complete. We assume the reader is familiar with Ritt-Wu's decomposition algorithm [RI1,WU2]. The following notions are needed frequently in this section.

For two polynomial sets ES, IS in $Q[x_1, ..., x_n]$ and an extension K of Q, let

$$Zero(ES) = \{x = (x_1, ..., x_n) \in K^n \mid \forall P \in ES, P(x) = 0\}$$

$$Zero(ES/IS) = Zero(ES) - \bigcup_{d \in IS} Zero(d).$$

We use $RZero(ES/IS)$ and $CZero(ES/IS)$ to denote $Zero(ES/IS)$ when K is $\mathbf{R}$ or $\mathbf{C}$ respectively. For an ascending chain (depending on context, sometimes maybe a triangular set) ASC and a polynomial P, let $prem(p, ASC)$ be the *pseudo remainder* of P wrpt ASC. For an ascending chain, we use the following notation:

$$PD(ASC) = \{P \in Q[X] \mid prem(P, ASC) = 0\}$$

Ritt-Wu's Zero Decomposition Theorem. For two finite polynomial sets ES and IS, we can either detect the emptiness of $Zero(ES/IS)$ or furnish a decomposition of the following form

$$Zero(ES/IS) = \cup_{i=1}^{l} Zero(PD(ASC_i)/IS)$$

where (a) for each $i \leq l$, ASC_i is irreducible such that $prem(P, ASC_i) \neq 0$ for $\forall P \in IS$; (b) there are not $i \neq j$ such that $PD(ASC_i) \subset PD(ASC_j)$.

We call such a decomposition an *irredundant decomposition*, and $Zero(PD(ASC_i)/IS)$ the *irreducible components* of $Zero(ES/IS)$.

4.1. The General Case

Definition 4.1. An algebraic statement (ES, IS, C) is called trivially true if its hypothesis is contradict, i.e, $CZero(ES/IS)$ is empty.

Definition 4.2. An algebraic statement $S = (ES, IS, C)$ is called unmixed if S is not trivially true and $Zero(ES/IS)$ is an unmixed quasi-variety, i.e.,

$$Zero(ES/IS) = \cup_{1 \leq i \leq k} Zero(PD(ASC_i)/IS)$$

where ASC_i are irreducible ascending chains with the same parameter set $\{u_1, ..., u_p\}$.

The variable set $\{u_1, ..., u_p\}$ in Definition 4.2 is called a *parameter set* of the unmixed statement.

If point P is introduced by a construction C, a *parameter set of* C consists of

two of the variables in a coordinate of P, if C is Construction 1;
one of the variables in a coordinate of P, if C is Construction 2 or 3;
$\emptyset$, if C is Construction 4 or 5 or 6.

For a statement of constructive type, the union of all the parameter sets of the constructions in the construction sequence of the statement is called a *parameter set of the statement*. Let n_1, n_2, n_3 be the numbers of Constructions 1-3 occurring in the construction sequence of the statement respectively. Then each parameter set of the statement has $d = 2n_1 + n_2 + n_3$ variables. We define d to be the *dimension of the statement*. Note that in a statement of constructive type, the variables in a set of parameters can take arbitrary values and, once their values are fixed, other variables can be generally determined by the geometric hypothesis. It is in this sense that we call them parameters.

The following theorem is the main result of this subsection and the completeness of the non-degenerate conditions can be deduced from this theorem.

Theorem 4.3. Let $S = (ES, IS, C)$ be the algebraic version of a statement of constructive type and $\{u_1, ..., u_q\}$ be a parameter set for the statement. Then we can decide in a finite number of steps whether S is trivially true or S is an unmixed statement with the u as its parameter set.

The sufficiency of the non-degenerate conditions generated by the constructions can be described in two aspects. First, they are the conditions under which the configurations of the hypothesis of the statements are normal. For example, when we take a point on a straight line we assume the line is well-defined; when we take the intersection of two lines we assume the two lines are well defined and they have a normal intersection. For the second aspect of the sufficiency, we first have

Theorem 4.4. If P dose not vanish on any irreducible component of $Zero(ES/IS)$, then for any polynomial C, the algebraic statement $S = (ES, IS, C)$ is universally true iff the new statement $S' = (ES, IS \cup \{P\}, C)$ is universally true.

Proof. Let $Zero(ES/IS) = \cup_{1 \leq i \leq k} Zero(PD(ASC_i)/IS)$ be an irredundant decomposition of $Zero(ES/IS)$. S is universally true iff $prem(C, ASC_i) = 0$ for $i = 1, ..., k$. Since $prem(P, ASC_i) \neq 0$ for $i = 1, ..., k$, we have the following irredundant decomposition

$$Zero(ES/IS \cup \{P\}) = \cup_{1 \leq i \leq k} Zero(PD(ASC_i)/IS \cup \{P\})$$

Then S' is universally true iff $prem(C, ASC_i) = 0$, $i = 1, ..., k$, i.e., iff S is universally true. ∎

For a statement S of constructive type, by Theorem 4.3 all the components are normal in the sense that they are with a parameter set of the statement as their parameter set. Then Theorem 4.4 implies that if $S = (ES, IS, C)$ is not universally true, a new statement $S = (ES, IS \cup \{P\}, C)$ (P is a polynomial) can not be universally true if $P \neq 0$ does not delete any normal component.

To prove Theorem 4.3, we first have

Lemma 4.5. Let $S = (ES, IS, C)$ be the algebraic version of a statement of constructive type and $\{u_1, ..., u_q\}$ be a parameter set for the statement, then either S is trivially true or we can furnish a decomposition of the following form:

$$Zero(ES/IS) = Zero(PD(ASC)/IS) \tag{4.6}$$

where (a) ASC is a (weak) ascending chain with the u as its parameter set; (b) $ASC \subset Ideal(ES)$; (c) No pseudo-remainders of the polynomials in IS wrpt ASC are zero.

Proof. The proof is a little bit lengthy. For details see Lemmas 4.5, 4.9 in [CG3]. ∎

Proof of Theorem 4.3. By Theorem 4.5, if the statement is not trivially true we get a decomposition like (4.6). Now the theorem comes from Theorem 2.6 in [GC1], since (4.6) is still true when $PD(ASC)$ is replaced by $QD(ASC)$. ∎

Remark. To confirm a statement of constructive type (ES, IS, C), we can use Lemma 4.5 to get a decomposition (4.6). If $prem(C, ASC) = 0$ the statement is universally true. To do this we do not need any polynomial factorization.

4.2. The Irreducible Case

Definition 4.7. An algebraic statement (ES, IS, C) is called irreducible if $Zero(ES/IS)$ $=$

$Zero(PD/IS)$ where PD is a prime ideal and no polynomials in IS belong to PD.

The non-degenerate conditions IS of an irreducible statement (ES, IS, C) is sufficient in the sense that, if the statement is not universally true, it cannot be universally true by adding more non-degenerate conditions unless the statement is trivially true under these non-degenerate conditions. Precisely, we have

Theorem 4.8. If an irreducible algebraic statement (ES, IS, C) is not universally true, then for any polynomial P either $(ES, IS \cup \{P\}, C)$ is trivially true or $(ES, IS \cup \{P\}, C)$ is still not universally true.

Proof. Let $Zero(ES/IS) = Zero(PD/IS)$ where PD is a prime ideal. If $(ES, IS \cup \{P\}, C)$ is not trivially true, then $Zero(PD/IS \cup \{P\})$ is not empty, i.e., $P \notin PD$. Thus, $(ES, IS \cup \{P\}, C)$ is still not universally true since $C \notin PD$. ∎

For a geometry statement of constructive type, we have

Theorem 4.9. Let (ES, IS, C) be the algebraic version of a constructive statement. If ASC in (4.6) is irreducible, then (ES, IS, C) is an irreducible algebraic statement.

Proof. $PD(ASC)$ is a prime ideal for any irreducible ascending chain [WU2]. ∎

Theorem 4.10. Let $S = (ES, IS, C)$ be the algebraic version of a constructive statement whose construction sequence consists of only Constructions 1-3, then either S is trivially true or S is an irreducible statement.

Proof. See Theorem 4.14 in [CG3]. ∎

4.3. The Connection between the Two Approaches

In this section, we shall prove that for geometry statements of constructive type, the two approaches to dealing with non-degenerate conditions have connections.

Theorem 4.11. Let (ES, IS, C) be an unmixed algebraic statement with parameter set $u_1, ..., u_q$. If (ES, C) is generally true wrpt the u, then (ES, IS, C) is universally true.

Proof. Let

$$Zero(ES/IS) = \cup_{i=1}^{t} Zero(PD(ASC_i)/IS)$$

be the irredundant decomposition of $Zero(ES/IS)$. ASC_i are irreducible ascending chains with $u_1, ..., u_q$ as their parameter set. Since (ES, C) is generally true wrpt $u_1, ..., u_q$, there is a polynomial D of the u such that $Zero(ES) \subset Zero(DC)$. Thus for each $i \leq t$ $Zero(PD(ASC_i)/IS) \subset Zero(ES/IS) \subset Zero(DC)$. This implies $prem(DC, ASC_i) = 0$. Then $prem(C, ASC_i) = 0$ as D is a polynomial of the u. This means (ES, IS, C) is universally true. ∎

Corollary. Let (ES, IS, C) be the algebraic version for a geometry statement of constructive type. If (ES, C) is generally true wrpt a parameter set of the statement, then the statement (ES, IS, C) is universally true.

Proof. The result comes from Theorem 4.11 and Theorem 4.3. ∎

Theorem 4.12. Let $S = (ES, IS, C)$ be the algebraic version for a constructive statement S. If ASC in (4.6) is irreducible, then S is universally true iff (ES, C) is generally true wrpt a parameter set of the statement.

Proof. By Theorem 4.11, we need only to prove one direction. Let $u_1, ..., u_q$ be a parameter set of the statement, then by Theorem 4.9, we have

$$Zero(ES/IS) = Zero(PD(ASC)/IS)$$

where ASC is an irreducible ascending chain with the u as parameter set and no pseudo remainders of the polynomials in IS wrpt ASC are zero. As (ES, IS, C) is universally true, we have $prem(C, ASC) = 0$. This implies that there is a polynomial D of the u such that DC belongs to the ideal generated by ASC. By Lemma 4.5 (b), $Zero(ES) \subset Zero(ASC)$. Now we have $Zero(ES) \subset Zero(DC)$. ∎

We can benefit from Theorems 4.11 and 4.12 by the following fact: proving a statement to be generally true is faster than proving the same statement to be true under certain non-degenerate conditions.

5. Experiment Results

If a geometry statement can be described in projective geometry, then it is true in Euclidean geometry iff it is true in Riemann geometry. For example, the Pascal theorem for a circle (p8 of [CH2]) is true in Riemann geometry. Such statements need no proof. We have proved about twenty theorems which are not statements in projective geometry including the butterfly Theorem, Ceva theorem, and Menelus theorem, etc. Here are two examples which are used to illustrate the usefulness of the results given in this paper.

Example 1. Let ABC be a triangle such that $AC \equiv BC$. D is a point on AC; E is a point on BC such that $AD \equiv BE$. F is the intersection of DE and AB. Show $DF \equiv EF$.

This statement is of constructive type, and a construction sequence of the statement is

Take an arbitrary points A and B.	Construction 1.
C is on $B(AB)$.	Construction 2.
D is on $L(AC)$.	Construction 2.
$E = L(BC) \cap (B, AD)$.	Construction 5.
$F = L(AB) \cap L(DE)$.	Construction 4.

The conclusion is cong(D, F, F, E).

Thus by Section 3.1, in the real number case the example is equivalent to the geometry statement $(HS, DS, \text{cong}(D, F, F, E))$, where

$$\begin{aligned} HS &= \{\ \text{cong}(A,C,C,B), \text{coll}(D,A,C,), \text{cong}(B,E,A,D),\\ &\qquad \text{coll}(E,B,C,), \text{coll}(F,D,E), \text{coll}(F,A,B)\ \}.\\ DS &= \{A = B, A = C, \text{pole}(B,B,C), \text{para}(A,B,D,E)\} \end{aligned}$$

Let $A = (0,0,1), B = (0, x_1, x_2), C = (x_3, x_4, x_5), D = (x_6, x_7, x_8), E = (x_9, x_{10}, x_{11})$, $F = (0, x_{12}, x_{13})$*. By Section 3.1, the algebraic version of the statement is (ES, IS, g) where

$$ES = \{\|B\| - 1, \|C\| - 1, \|D\| - 1, \|E\| - 1, \|F\| - 1, h_1, h_2, h_3, h_4, h_5\}$$

$h_1 = x_5^2 - (x_1x_4 + x_2x_5)^2 = 0$	cong(A,C,C,B).
$h_2 = x_3x_7 - x_4x_6 = 0$	coll$(D,A,C,)$.
$h_3 = (x_1x_{10} + x_2x_{11})^2 - x_8^2 = 0$	cong(B,E,A,D).
$h_4 = x_1x_3x_{11} - x_2x_3x_{10} + (-x_1x_5 + x_2x_4)x_9 = 0$	coll$(E,B,C,)$.
$h_5 = (x_6x_{10} - x_7x_9)x_{13} + (-x_6x_{11} + x_8x_9)x_{12} = 0$	coll(F,D,E).

$IS = \{d_1, d_2, d_3, d_4\}$.

$d_1 = \|A - B\|\|A + B\| \neq 0$	$A \neq B$.
$d_2 = \|A \times C\| \neq 0$	$A \neq C$.
$d_3 = \|(B \times C) \times B\| \neq 0$	$\neg$pole(B,B,C).
$d_4 = \|(A \times B) \times (D \times E)\| \neq 0$	$\neg$para(A,B,D,E).

* Here the coordinates of a point, say $C = (x_3, x_4, x_5)$, actually means $C = \pi(x_3, x_4, x_5)$. The same is true for Example 2.

$$g = (x_{11}^2 - x_8^2)x_{13}^2 + ((2x_{10}x_{11} - 2x_7x_8)x_{12})x_{13} + (x_{10}^2 - x_7^2)x_{12}^2 = 0 \quad \text{cong}(D, F, F, E).$$

By Theorem 2.7, the statement can be reduced to one substatement (ES', Is, g) where $ES' = \{\|B\| - 1, \|C\| - 1, \|D\| - 1, \|E\| - 1, \|F\| - 1, h_1', h_2, h_3', h_4, h_5\}$, and

$$h_1' = x_5 - (x_1x_4 + x_2x_5) = 0 \qquad \text{congl}(A, C, C, B).$$
$$h_3' = (x_1x_{10} + x_2x_{11}) - x_8 = 0 \qquad \text{congl}(B, E, A, D).$$

To prove the statement using Ritt-Wu's decomposition algorithm, we first have the following decomposition: $Zero(ES'/IS) = Zero(PD(ASC_1)/IS)$ where

$ASC_1 = x_2^2 + x_1^2 - 1$, $(x_2^2 - 2x_2 + x_1^2 + 1)x_4^2 + (x_2^2 - 2x_2 + 1)x_3^2 - x_2^2 + 2x_2 - 1$, $(x_2 - 1)x_5 + x_1x_4$, $x_3x_7 - x_4x_6$, $x_8^2 + x_7^2 + x_6^2 - 1$, $x_9^2 - x_6^2$, $((x_2^2 + x_1^2)x_3)x_{10} + (x_1x_2x_5 - x_2^2x_4)x_9 - x_1x_3x_8$, $x_2x_{11} + x_1x_{10} - x_8$, $(x_6^2x_{11}^2 - 2x_6x_8x_9x_{11} + x_6^2x_{10}^2 - 2x_6x_7x_9x_{10} + (x_8^2 + x_7^2)x_9^2)x_{12}^2 - x_6^2x_{10}^2 + 2x_6x_7x_9x_{10} - x_7^2x_9^2$, $(x_6x_{10} - x_7x_9)x_{13} + (-x_6x_{11} + x_8x_9)x_{12}$.

We have $prem(g, ASC_1) = 0$ which means the statement (ES, IS, g) is universally true.

Example 2 (Ceva Theorem). For a triangle ABC and a point G, let AG, BG, CG intersect BC, AC, AB in F, E, D respectively. Show that:

$$\frac{\sin(AD)\sin(BF)\sin(CE)}{\sin(DB)\sin(FC)\sin(EA)} = 1$$

Strictly speaking, this example is not a geometry statement according to Definition 2.1. To describe the statement, we need the following new predicates:

(a) cos-dis(A, B) equals: $\langle A, B\rangle^2$,

(b) sin-dis(A, B) equals: $1 - \langle A, B\rangle^2$,

and we need to assume that the conclusion of a geometry statement can be any polynomial equation.

The construction sequence of the statement is

Take arbitrary points G, A, B, and C in **P2**.	Construction 1.
$D = L(GC) \cap L(AB)$.	Construction 4.
$E = L(GB) \cap L(AC)$.	Construction 4.
$F = L(GA) \cap L(BC)$.	Construction 4.

The conclusion: sin-dis(AD)sin-dis(BF)sin-dis(CE) - sin-dis(DB) sin-dis (FC) sin-dis (EA)=0.

Now the statement can be proved in the same way as Example 1.

References.

[CH1] S.C. Chou, Proving and Discovering Geometry Theorems Using Wu's Method, PhD Thesis, Dept. of Math., University of Texas at Austin, 1985.

[CH2] S.C. Chou, *Mechanical Geometry Theorem Proving,* D.Reidel Publishing Compan
1988.

[CG1] S.C. Chou and X.S. Gao, Ritt-Wu's Decomposition Algorithm and Geometry Theo rem Proving, *10th International Conference on Automated Deduction*, M.E. Stick (Ed.) pp 207–220, Lect. Notes in Comp. Sci., No. 449, Springer-Verlag, 1990.

[CG2] S.C. Chou and X.S. Gao, A Class of Geometry Statements of Constructive Typ and Geometry Theorem Proving, TR-89-37, Computer Sciences Department, Th University of Texas at Austin, November 1989, to appear in CADE'92.

[CG3] S.C. Chou and X.S. Gao, Mechanical Theorem Proving in Riemann Geometr TR-90-03, Computer Sciences Department, The University of Texas at Austir February, 1990.

[CK1] S.C. Chou and H.P. Ko, On the Mechanical Theorem Proving in Minkowskian Plan Geometry, *Proc. of Symp. of Logic in Computer Science*, pp187-192, 1986.

[CY1] S.C. Chou and Yang Jingen, On the Algebraic Formulation of Certain Geometr Statements and Mechanical Geometry Theorem Proving, *Algorithmica*, Vol. 4, 198 237-262.

[GA1] X.S. Gao, Transcendental Functions and Mechanical Theorem Proving in Elemen tary Geometries, *Journal of Automated Reasoning*, 6:403-417, 1990, Kluwer Aca demic Publishers.

[GA2] X.S. Gao, Mechanical Theorem Proving In Caley-Klein Geometries *J. of Sys. Sc and Math. Sci.*, 1992, 3: 260–273.

[GC1] X.S. Gao & S.C. Chou, On the Dimension of an Arbitrary Ascending Chain, t appear in *Chinese Science Bulletin.*

[KA1] D. Kapur, Geometry Theorem Proving Using Hilbert's Nullstellensatz, *Proc. o SYMSAC'86*, Waterloo, 1986, 202–208.

[KU1] B.A. Kutzler, Algebraic Approaches to Automated Geometry Theorem Proving Phd Thesis, Johnnes Kepler University, 1988.

[RI1] Ritt, J.F., *Differential Algebra*, Amer. Math. Soc., 1954.

[RY1] P. Ryan, *Euclidean and Non-Euclidean Geometries*, Cambridge, 1986.

[WG1] D.M. Wang & X.S. Gao, Geometry Theorems Proved Mechanically Using Wu' Method, Part on Elementary Geometries, MM preprint No2 1987.

[WU1] Wu Wen-tsün, On the Decision Problem and the Mechanization of Theorem in Ele mentary Geometry, *Scientia Sinica 21(1978)*, 159–172; Re-published in *Automate Theorem Proving: After 25 years*, A.M.S, Contemporary Mathematics, 29(1984) 213–234.

[WU2] Wu Wen-tsün, Basic Principles of Mechanical Theorem Proving in Geometries, Vol ume I: Part of Elementary Geometries, Science Press, Beijing (in Chinese), 1984.

[WU3] Wu Wen-tsün, Toward Mechanization of Geometry — Some Comments on Hilbert'

"Grundlagen der Geometrie", *Acta Math. Scientia*, 2(1982), 125–138.

[YA1] I.M. Yaglom, *A Simple Non-Euclidean Geometry and Its Physical Basis*, Springer-Verlag, 1979.

Appendix A. Some Basic Properties of P2

In this appendix, we shall prove some basic properties of **P2** using our prover based on Wu's method. The results given in this section provide proofs for the equivalence of the inequation part DS in Section 3.1 to certain combination of predicates in the real number case. For example, in case 4.3, $DS = \{\|p_1 \times (p_2 \times (p_3 \times p_4))\| \neq 0\}$ which is equivalent to $p_1 \times (p_2 \times (p_3 \times p_4)) \neq 0$ in the real number case. Now by $A12$ below, we get the result in Section 3.1.

Let p_i be points on **S2** such that $\pi(p_i) = P_i \in \mathbf{P2}$ for $i = 1, ..., 6$.

A1. $P_1 = P_2$ iff $(p_1 \times p_2) = 0$.

Proof. If $P_1 = P_2$, we have $p_1 = p_2$ or $p_1 = -p_2$. Thus $(p_1 \times p_2) = 0$. If $(p_1 \times p_2) = 0$, there is a number r such that $p_1 = rp_2$. Then $\|p_1\| = r^2\|p_2\| = r^2 = 1$. We have $p_1 = p_2$ or $p_1 = -p_2$, i.e., $P_1 = P_2$. ∎

A2. Para(P_1, P_2, P_3, P_4) iff $P_1 = P_2$ or $P_3 = P_4$ or (coll(P_3, P_1, P_2) and coll(P_4, P_1, P_2) and coll(P_1, P_3, P_4) and coll(P_2, P_3, P_4)).

Proof. By the formulas

$$(a \Rightarrow (b \text{ or } c)) \longleftrightarrow ((a \text{ and } \neg b) \Rightarrow c)$$
$$((a \text{ or } b) \Rightarrow c) \longleftrightarrow ((a \Rightarrow c) \text{ and } (b \Rightarrow c))$$
$$((a \Rightarrow b) \text{ and } (a \Rightarrow c)) \longleftrightarrow (a \Rightarrow (b \text{ and } c))$$

B2 can be reduced to the following geometry statements:

$$(\text{para}(P_1, P_2, P_3, P_4) \wedge P_1 \neq P_2 \wedge P_3 \neq P_4) \Rightarrow \text{coll}(P_3, P_1, P_2)$$
$$(\text{para}(P_1, P_2, P_3, P_4) \wedge P_1 \neq P_2 \wedge P_3 \neq P_4) \Rightarrow \text{coll}(P_4, P_1, P_2)$$
$$(\text{para}(P_1, P_2, P_3, P_4) \wedge P_1 \neq P_2 \wedge P_3 \neq P_4) \Rightarrow \text{coll}(P_1, P_3, P_4)$$
$$(\text{para}(P_1, P_2, P_3, P_4) \wedge P_1 \neq P_2 \wedge P_3 \neq P_4) \Rightarrow \text{coll}(P_2, P_3, P_4)$$
$$(P_1 = P_2) \Rightarrow \text{para}(P_1, P_2, P_3, P_4)$$
$$(P_3 = P_4) \Rightarrow \text{para}(P_1, P_2, P_3, P_4)$$
$$(\text{coll}(P_3, P_1, P_2) \wedge \text{coll}(P_4, P_1, P_2) \wedge \text{coll}(P_1, P_3, P_4) \wedge \text{coll}(P_2, P_3, P_4)$$
$$\wedge P_1 \neq P_2 \wedge P_3 \neq P_4) \Rightarrow \text{para}(P_1, P_2, P_3, P_4)$$

which can be proved by our prover (Theorem 2.3). ∎

We have proved the following statements similarly.

A3. Pole(P_1, P_2, P_3) iff perp(P_1, P_2, P_2, P_3) and perp(P_1, P_3, P_2, P_3).

A4. Pole(P_1, P_1, P_2) iff $P_1 = P_2$.

A5. Para(P_1, P_2, P_1, P_3) iff coll(P_1, P_2, P_3).

A6. We have $(p_1 \times p_2) \times (p_3 \times (p_4 \times p_5)) = 0$ iff pole(P_3, P_4, P_5) or (perp(P_1, P_2, P_4, P_5) and coll(P_1, P_2, P_3)).

A7. We have $(p_1 \times p_2) \times (p_3 - p_4) = 0$ or $(p_1 \times p_2) \times (p_3 + p_4) = 0$ iff $P_1 = P_2$ or $(\mathrm{cong}(P_1, P_3, P_1, P_4)$ and $\mathrm{perp}(P_1, P_2, P_3, P_4))$.

A8. We have $(p_1 \times p_2) \times (p_1 - p_2) = 0$ or $(p_1 \times p_2) \times (p_1 + p_2) = 0$ iff $P_1 = P_2$.

A9. We have $(p_1 \times (p_2 \times p_3)) \times (p_4 \times (p_5 \times p_6)) = 0$ if and only if $\mathrm{pole}(P_1, P_2, P_3)$ or $\mathrm{pole}(P_4, P_5, P_6)$ or $(P_1 = P_2$ and $\mathrm{cperp}(P_1, P_2, P_3, P_5, P_6))$ or $(\mathrm{perp}(P_1, P_4, P_2, P_3)$ and $\mathrm{perp}(P_1, P_4, P_5, P_6))$.

A10. We have $(p_1 \times (p_2 \times p_3)) \times (p_4 \pm p_5) = 0$ iff $\mathrm{pole}(P_1, P_2, P_3)$ or $(\mathrm{cong}(P_1, P_4, P_1, P_5)$ and $\mathrm{cperp}(P_1, P_2, P_3, P_4, P_5))$.

A11. We have $(p_1 \pm p_2) \times (p_3 \pm p_4) = 0$ iff $P_1 = P_2$ or $P_3 = P_4$ or $(\mathrm{cong}(P_1, P_3, P_2, P_4)$ and $\mathrm{cong}(P_1, P_4, P_3, P_2))$.

A12. We have $o \times (p_1 \times (p_2 \times p_3) = 0$ iff $\mathrm{pole}(P_1, P_2, P_3)$ or $(\mathrm{coll}(O, P_2, P_3)$ and $\mathrm{polar}(O, P_1))$.

A13. We have $o \times (p_1 \pm p_2) = 0$ iff $(\mathrm{coll}(O, P_1, P_2)$ and $\mathrm{cong}(O, P_1, O, P_2))$.

Author Index